RADIATION
PROTECTION
DOSIMETRY

A RADICAL REAPPRAISAL

RADIATION PROTECTION DOSIMETRY

A RADICAL REAPPRAISAL

JACK A. SIMMONS
Professor of Radiation Biophysics
University of Westminster
London, England

and

DAVID E. WATT
Honorary Reader in Radiation Biophysics
University of St. Andrews
St. Andrews, Scotland

Medical Physics Publishing
Madison, Wisconsin

Library of Congress Cataloging-in-Publication Data

Simmons, Jack A., 1934–
     Radiation protection dosimetry: a radical reappraisal / Jack A. Simmons
and David E. Watt.
          p.   cm.
     Includes bibliographical references and index.
     ISBN 0-944838-87-1 (softcover : alk. paper)
     1. Radiation dosimetry.   2. Radiation--Safety measures.
3. Radiation--Physiological effect.   I. Watt, D. E. (David Edwin)
II. Title.
     [DNLM: 1. Radiometry.   2. Radiation Protection.   3. Relative
Biological Effectiveness.   4. DNA--radiation effects.   WN 660 S592r 1999]
RA569.S52   1999
612'.01448--dc21
DNLM/DLC
for Library of Congress   98-52973
                                                                         CIP

Medical Physics Publishing
4513 Vernon Blvd.
Madison, WI 53705-4964
608-262-4021

Information in this book is provided for instructional use only. Because of the possibility for human error and the potential for change in the medical sciences, the reader is strongly cautioned to verify all information and procedures with an independent source before use. The authors, editors, and publisher cannot assume responsibility for the validity of all materials or for any damage or harm incurred as a result of the use of this information.

Printed in the United States of America

Contents

Chapter 5. Radiation as a Probe for the Physical Investigation of Radiosensitive Structure in Biological Targets

Contents

List of Figures

Figure

Figure

Figure

Figure

List of Tables

Table

Preface

Attempts at quantifying the "amount" of radiation began almost immediately after the discovery of X-rays in 1895. Methods based on the phenomenon of ionization in air proved to be the easiest and most reliable to implement, and these became the accepted methods of quantifying amounts of radiation when the international scientific community began to develop standards in the 1920s. Twenty years later, by the late 1940s, however, the drawbacks and limitations of measurements based on ionization became apparent. Thinking switched to the concept of energy absorbed per unit mass of material irradiated as the fundamental measure, even though this could generally not be measured. All recommendations by both the International Commission on Radiation Units and Measurements (ICRU) and the International Commission on Radiological Protection (ICRP) since 1953 have therefore been standardized in these terms.

The dosimetry ideas promulgated by the ICRU and the ICRP have gained a wide measure of acceptance. Nevertheless, in recent years an increasing number of dissenting voices have been raised, mostly in barroom discussions after meetings, but more and more in scientific conferences and in the open, peer-reviewed literature. Stimulated by such publications and discussions, in this monograph the authors are attempting a radical reappraisal of radiation protection dosimetry; fundamentally, the idea of "energy absorbed per unit mass" as a basic unit and the quantities derived from it will be challenged. We hope that as a result a simpler, more logical system of radiation units and measurements can be created. We also hope that use of these new ideas will allow a more direct connection to be made between the amount of radiation (of all types) and the effects caused by the different amounts, especially at low doses. If these hopes are fulfilled, a revolutionary development in radiation science will have taken place.

Acknowledgments

It is impossible in a monograph such as this to acknowledge formally all those people who, directly or indirectly, have made its publication possible. Where appropriate, references to the open scientific literature have been given in the usual manner. Elsewhere, we have indicated the points at which we have paraphrased (we hope fairly!) the ideas of the International Commission on Radiological Protection (ICRP) and the International Commission on Radiation Units and Measurements (ICRU). However, many of our ideas will have arisen as a result of discussions with colleagues over many years, and we take this opportunity of expressing our thanks to them.

We also wish to thank our respective institutions for allowing us the time to undertake the writing of this work. In addition, the fact that they have not tried to deter us from putting forward some very controversial views is greatly appreciated.

Finally, our thanks must go to Prof. John Cameron, at whose suggestion this book was undertaken, and the staff at Medical Physics Publishing for their help and support.

1.

Introduction—A Historical Precedent in Astronomy

Throughout scientific history it has become necessary, at times, to undertake a fundamental reconsideration of the basic ideas involving accepted concepts. Such a re-thinking is essential when inconsistencies between theory and experiment become too glaring to be ignored, or when a very clumsy theory can be replaced by a simpler, more elegant one. A prime example of this can be found in the orbits of the planets in astronomy and, for reasons given later, we start with a brief history of this subject.

After the early work of Plato, Aristotle, and Eratosthenes in the third and fourth centuries B.C., major advances in astronomy were made by Hipparchus in the first century B.C. Unfortunately, most of his writings have been lost, so that our knowledge of his contributions is derived from his disciple Claudius Ptolemaeus (Ptolemy), who lived about 3 centuries later. From his observations (and probably those of others) Hipparchus realized that the Sun's apparent motion in the ecliptic is not uniform but varies systematically during the year. It was generally believed that the Earth was stationary and that everything revolved around it. Hipparchus believed in the perfectness of uniform circular motion, and therefore offered as one explanation the idea of eccentricity. Quite simply, the Earth from which he was observing was not exactly at the center of the circle described by the Sun. Calculations based on this idea were consistent with observations of the Sun's motion to within their limits of accuracy.

An alternative explanation, more fully developed by Ptolemy, was a system of deferents and epicycles. In this system, deferents were large circles centered on the Earth, and epicycles were small circles whose centers moved

I

around the circumferences of the deferents. The Sun, the Moon, and the planets moved around the circumference of their own epicycles.

Unfortunately, with the larger number of observations available to him, even this complex idea did not enable Ptolemy to account for all the astronomical measurements at that time. He therefore had to introduce yet another concept—that of the equant. This was an imaginary point placed on the diameter of each deferent, at a position opposite to that of the Earth from the center of the deferent. Ptolemy then specified that the motion of the center of each planet's epicycle should appear uniform when viewed from the extant. Finally, to account for the deviations of the planets from the ecliptic, Ptolemy postulated that the planes of the deferents were tilted up so that they were inclined to the ecliptic at various small angles.

Although this system enabled astronomers to account for many phenomena and to make predictions of movement, the improvements in the accuracy of the observations during the succeeding centuries resulted in increasing discrepancies with the theory. Hence, by about 1500 doubts were being expressed about the soundness of the Ptolemaic ideas, most notably by Copernicus. He realized that the many circles and their displacements from the center of the Earth were undermining the simple foundations of the generally accepted theory and set about finding a simpler, more realistic explanation.

Copernicus made comparatively few observations—only 27 are recorded in his own notebooks, although there are quite possibly a few more elsewhere. He was, however, a highly skilled mathematician and, perhaps more importantly, appreciated the significance of the concept of "relative motion." Therefore, when Copernicus found references to the motion of the Earth in early Greek writings, he was able to incorporate such hypotheses into his own thinking. He proposed that the apparent daily motions of the stars, the annual motion of the Sun, and the movements of the planets result from the Earth's daily rotation on its axis and yearly revolution around the Sun. Hence, in contradistinction to the Ptolemaic system, it was the Sun, not the Earth, that was stationary at the center of the universe. Correspondingly, the other planets also revolved around the Sun, with Mercury (the innermost) having a period of about 88 days, and Saturn (then the outermost known), a period of about 30 years. Only the Moon retained its status of revolving around the Earth.

The Copernican system quickly appealed to a large number of independent-minded astronomers and mathematicians. Its attraction lay not only in its elegance but also, in part, to the fact that it broke with traditional doctrines. It provided a realistic alternative to the Ptolemaic view of the universe and, by implication, rejected the Aristotelian notion of the Earth at the center. Both these ideas had been raised almost to the level of religious dogma and, like religious dogma, their continued acceptance was based more on the supposed authority of their originators rather than on firmly supported clear thinking. In breaking with tradition, Copernicus provided the platform from which the revolutionary ideas of the following century could be developed.

It is the authors' contention that, corresponding to the fundamental change brought about by Copernicus in the field of astronomy, the time is ripe for a fundamental change in the field known as radiation dosimetry, especially as it applies to biological effects. In most cases it is not the fundamental quantity (energy absorbed) that is measured but a (hopefully) related one—ionization in a gas. Various manipulations are performed on the measurements to calculate a "dose" (reminiscent of Ptolemy's epicycles). Doses are then "weighted" to give "dose equivalent" (tilting the planes of the ecliptic?), and the values of the "weights" are not known to any realistic degree of accuracy (the lost days between the Julian and Gregorian calendars). The whole radiation protection dosimetry system is given a spurious look of conviction by various national and international commissions, many of whose long-term members are, understand-ably, reluctant to change their views. Fuller details of the history and weaknesses of the present system are given in Chapter 2.

2.

A Brief Early History of Radiation Units and Radiation Protection

Just as Chapter 1 did not detail the full history of early astronomy, this chapter will not detail the full history of radiation units or the full history of recommendations for protection. Nevertheless, it is pertinent to give a brief account of the developments leading to the present situation.

Within a couple of years of the discovery of X-rays by Wilhelm Conrad Roentgen in 1895, in Paris Jean Baptiste Perrin had constructed an early form of what we would now call a free-air chamber. With this he was able to measure what he called the "quantity" of roentgen rays, preferring this term to the possible alternative, "intensity." A colleague of Perrin's, Paul Villard, subsequently improved the experimental arrangement (in effect, virtually converting it into a digital meter!) and proposed that the ionization he was measuring should be the basis for the unit of quantity for X-rays.

Other physicochemical effects were also being used to measure quantities of X-rays. The blackening of photographic film that Roentgen observed was one such method. Another was the change in color of certain chemicals (e.g., the "pastille unit" for barium platino-cyanide capsules or the "Holtzknect" for mixtures of potassium chloride and sodium carbonate). However, their dependence on personal estimation of color changes, the influence of the change in wavelength of the X-rays, and the need to employ reagents that varied in quality and sensitivity, combined to make such measuring systems unsuitable for anything except rough approximations of dose measurements.

As a result, the ionization measuring instrument and the unit of dose defined in physical terms (i.e., the quantity of charge liberated by the radiation in a defined volume of air) became generally accepted in the 1920s as a means of measurement that could be more accurately reproduced than with any other

5

system. The problem remained, however, to relate this to its effect on the human body; what, really, is the biological unit? For many years this was the Unit Skin Dose (USD), which was agreed to be the amount of radiation that would create a clear but reversible erythema of the irradiated skin. Although the USD was steadily abandoned after the adoption in 1928 of the roentgen (r) as the unit of dose, the underlying question of biological effect remains to be properly answered.

The harmful effects of X-rays on tissue were recognized not long after their discovery. As a result, recommendations were formulated by those working in the field for protection against such effects and also against the potential hazards of working with high-voltage electrical equipment. Originally, these recommendations concerned working hours, the physical design of X-ray departments, and the thickness of lead shielding for different tube voltages. Subsequently, in 1934 the International Commission on Radiological Protection (ICRP) suggested that under satisfactory working conditions a person in normal health could tolerate an exposure to X-rays of approximately 0.2 r per day; in this context, "tolerate" meant that injuries to the superficial tissues or changes in the blood were unlikely to occur. [In the United States, the National Commission for Radiation Protection (NCRP) adopted a level of 0.1 r per day in 1934.] This concept of a "tolerance dose" remained unchanged in the revised recommendations of the ICRP meeting held in 1937.

Developments in nuclear physics and their practical applications between 1938 and 1950 greatly increased the scope of potential hazards. With the production of nuclear weapons and the operation of nuclear reactors, the potential dangers arising from neutrons became of much greater significance. At the same time biological research led to an extension of knowledge of the dangers associated with ionizing radiations. This increase not only brought about a realization of the importance of certain effects, particularly carcinogenic and genetic ones, but also provided more information as to the permissible levels of radiation. As a result, the ICRP adopted new radiation safety standards with more rigid criteria at its sixth meeting in 1950.

At this meeting, it was agreed that it was fundamental to express whole-body exposure in terms of a single number. The ICRP recognized, however, that it was not practicable to express the maximum permissible hazards in terms of a single parameter. It was, for example, considered highly desirable to derive values of maximum permissible concentrations of radioactive materials both in the air and in drinking water. The ICRP further decided that the previously accepted value of 1 r per week for maximum permissible exposure to external radiation needed revision. The ICRP also noted the added difficulty that the roentgen was not an acceptable unit of dose for all types of ionizing radiation.

The values proposed for maximum permissible exposures were those that could be regarded as posing hazards which were small compared to others in daily life. However, the figure of 1 r per week previously adopted appeared very close to the threshold for adverse effects, particularly for radiations of high energy that were being encountered more frequently than in the past. It was therefore recommended that, for whole-body exposure to X-, gamma, and beta radiation, the maximum permissible dose received by the surface of the body should be 0.5 r per week, a dose that corresponded to 0.3 r in free air. (The terms "exposure" and "dose" seem to be used interchangeably in this ICRP report [ICRP, 1950].) All of the linked recommendations were based on considerations of the equivalent energy absorbed in tissue coupled with the appropriate relative biological efficiency (RBE).

The RBE of any given radiation was defined by comparison with the gamma radiation from radium filtered by 0.5 millimeters (mm) of platinum. It was expressed numerically as the inverse of the ratio of the doses of the two radiations (in ergs per gram of tissue) required to produce the same biological effect under the same conditions. For purposes of calculation, it was assumed that the RBE of a given radiation is the same for all effects, with the single exception of gene mutations.

The ICRP also acknowledged that, at that time, it was not in a position to make firm recommendations regarding the maximum permissible amounts of radioactive material that could be taken into, or retained in, the body. Nevertheless, it was felt to be appropriate to draw attention to data on maximum permissible exposures to a small number of radionuclides for occupational workers.

For the alpha-particle emitters ^{226}Ra, ^{239}Pu, and ^{210}Po, the body burden limits were 0.1, 0.04, and 0.005 microcurie (μCi), respectively. It is interesting to note that these values were proposed on the basis of a mixture of clinical data (for radium in man) and animal experiments; they did not arise from the use of the proposed RBEs for alpha particles. For the beta-particle emitters, values were calculated so as to give the dose equivalent to 0.3 r per week to the whole body.

The rationale behind the recommended values of RBE has never been explained (10 for fast neutrons and 20 for alpha particles). Very few experimental data were available; Douglas Lea, in his classic book *Actions of Radiations on Living Cells* [Lea, 1946], gave a fairly detailed account of the various investigations which had been carried out up to that time. A summary of the relevant results shows the following:

- For mutations, the RBEs of fast neutrons and alpha particles were 0.66 and 0.29, respectively.

- For chromosome aberrations (often regarded as a precursor of malignant transformation), the RBE of neutrons ranged from 2.2 to 4.3 (with the exception of bean root tips, 6.0).
- For chromatid breaks, the RBEs of fast neutrons and alpha particles were 2.3 and 1.1, respectively.

It is therefore clear that the ICRP recommendations proposing that the RBE of alpha particles should be double that of fast neutrons are in direct contradiction of these results.

There is, of course, the possibility that early forms of neutron dosimetry were so wildly inaccurate that such results could be ignored. The ICRP committee did not, however, argue along these lines; indeed, it did not cite *any* evidence to justify its figures. Furthermore, it is interesting to note that modern observations are broadly in line with these values in Lea's book, essentially showing that the RBE does not reach double figures either for neutrons or for alpha particles.

The particular case of mutations is especially noteworthy. These appear to fall into the category of what has become known as "one-hit detectors." This term implies that one hit by an incident particle on a sensitive target will cause that target to undergo an irreversible change. Examples include the conversion of the ferrous to the ferric ion in the Fricke dosimeter and the induction of stable free radicals in alanine. For both of these, the relative effectiveness of fast neutrons is about 0.65 compared to gamma rays.

At the same time as the meeting of the ICRP, a meeting of the International Commission on Radiation Units and Measurements (ICRU) was also held—in fact there was cross-membership between the two commissions. The ICRU meeting proved to be something of a landmark, for it represented the "half-way house" between measurements of ionization in air and of energy absorbed in tissue. In the opening paragraph of its report [ICRU, 1950], the ICRU recommended that for the correlation of the dose of any ionizing radiation with its biological or related effects, the dose should be expressed in terms of the quantity of energy absorbed per unit mass (ergs per gram) of irradiated material at the place of interest. However, in a later paragraph the ICRU considered that the roentgen, in view of its long-established usefulness, should remain as the unit of X- and gamma-ray quantity or dose. It further recommended that when describing the conditions of X-ray treatment, a distinction should be made between the quantity of radiation measured in air and the quantity of radiation estimated to have been received by the tissue. Because the symbol "r" was reserved for the unit, the amount of the dose could be designated by the letter "D". For example, the dose measured in air might be D = 300 r. Then the dose measured at the surface of a body increased by backscatter might be D_o = 500 r and at some depth x, it might be D_x = 200 r. Readers were warned that D was not to be confused with the energy actually absorbed by the tissue.

Following this creation of the maximum imaginable degree of confusion concerning the term "dose," at its meeting in 1953 the ICRU was ready to make the definitive change from the quantity of radiation as measured by the roentgen to the absorbed dose as measured by the amount of energy imparted to matter per unit mass of irradiated material [ICRU, 1953]. The unit was to be the rad, with 1 rad equal to 100 ergs per gram. Despite this change, the roentgen remained the unit for X- and gamma-ray dose! A further improvement was made 3 years later when "dose" for X- and gamma rays was replaced by "exposure dose" and, simultaneously, the roentgen was slightly redefined.

At the same time as the 1953 ICRU meeting was taking place, once again some of the subcommittees of the ICRP were meeting. Unfortunately, their preliminary reports could not be agreed on, and the final ICRP recommendations were not published until 1955 [ICRP, 1955]. In harmony with the ICRU, these recommendations gave values of maximum permissible doses of X- and gamma radiations expressed in roentgens. However, to add to the confusion, they did express certain dosage values in terms of the rem (*r*oentgen *e*quivalent in *m*an), although the ICRU did not officially recognize this unit.

The rem was defined by the ICRP as the absorbed dose of any ionizing radiation that has the same biological effectiveness as 1 rad of X-radiation with an average specific ionization of 100 ion pairs per micron of water, in terms of its air equivalent, in the same region. A dose in rems was equal to the dose in rads multiplied by the appropriate RBE.

This definition was followed by a table of values of RBE listed as a function of both average specific ionization (ion pairs per micron of water) and average linear energy transfer (LET) in kiloelectron volts (keV) per micron. Apart from X-rays and electrons (LET 3.5 or less) whose RBE was defined as 1, all other values were given as ranges; e.g., specific ionizations of 650 to 1500 ion pairs per micron (corresponding to LETs of 23 to 53) carried RBE values of 5 to 10. From this table, one could deduce that an RBE in the range 10 to 20 should be assigned to alpha particles from naturally occurring emitters. However, another ICRP subcommittee was more specific; naturally occurring alpha particles were to be assigned an RBE of 10, the same value as that for fast neutrons. The reduction from the previous value of 20 was never explained.

Even more fundamentally, no explanation was offered for the values of RBE given in the table. This was a surprising omission because a year or so previously the U.S. National Bureau of Standards had published a full and detailed report by John Boag (No. 2946) [Boag, 1954] on the RBE of a wide range of different ionizing radiations. This report listed the results of over 150 investigations, and showed clearly that the values of RBE obtained depended critically on the endpoint chosen. The great majority of these values lay between 1 and 10; in particular, for late reactions in the tissues of cancer patients irradiated by fast neutrons, it was estimated that the RBE was about 5. Boag did not attempt to relate RBE to either specific ionization density or the LET. He presumably regarded this as a meaningless exercise, and it is therefore difficult to understand why the ICRP should have adopted this practice.

In 1959 the ICRU expressed misgivings about the use of the same term "RBE" in both experimental radiobiology and radiation protection. RBE values depend not only on the biological system studied but also on other variables such as absorbed dose rate, fractionation, and oxygen tension. For radiation protection purposes, certain values of RBE had been selected from the literature and promulgated by the ICRP in 1955. This was clearly unsatisfactory and, at a joint meeting in 1962 of representatives of both the ICRP and the ICRU, it was agreed that the term "RBE" be used only in radiobiology and that another term be used for the LET-dependent factor applicable for radiation protection purposes [ICRU 10a, 1962]. As before, the absorbed doses were to be multiplied by this factor to obtain a quantity that expressed on a common scale for all ionizing radiations the irradiation incurred by exposed persons. The name recommended for this factor was "quality factor" (QF, later shortened to Q).

Provision for other modifying factors was also made. Thus, a "distribution factor" (DF) could be introduced to express the modification of the biological effect due to nonuniform distribution of internally deposited isotopes. The product of absorbed dose and modifying factors was to be termed the "dose equivalent" (DE), i.e.,

$$DE = D \cdot (QF) \cdot (DF) .$$

The unit of dose equivalent was the rem.

The actual values of QF were to be the same as the RBE values previously recommended. These values were now considered to be solely a function of LET; i.e., the link to average specific ionization was quietly dropped. In this context, the linear energy transfer was to be LET_∞ (the "Stopping Power") so that delta rays were not counted as separate tracks. Some simplification was allowed; for example, a single value of QF = 10 was recommended for all fast neutrons. However, recognizing that the lens of the eye was especially sensitive to particulate radiation of high LET, the ICRP recommended that a value of QF = 30 should be used instead of 10.

The idea that delta rays need not be counted as separate tracks is questionable. Particles of different mass and charge may, for suitable energies, have the same average rate of energy loss, but will give rise to differences in the delta-ray distribution. Where the latter make a significant contribution to the effect studied, clearly the LET_∞ is not an appropriate parameter.

One important comment in the ICRP Report of 1962 is often overlooked. It was noted that there are certain radiation exposure conditions where the QF concept as outlined previously was either inapplicable or could only be applied with major qualifications. Meaningful examples are those where gross nonuniformity of dose distribution occurs, as with bone-seeking radioactive nuclides or with radioactive particles in the lung. For the bone-seekers, special methods were developed that involved the use of an additional "relative damage factor." In the case of the lung, no such special methods could be used because of lack of relevant information; the comment was therefore made that "an estimate of DE to the critical tissue determined merely by the product of QF and the mean dose might well be greatly in error" [ICRP 6, 1962].

All of these recommendations remained broadly unchanged when the ICRP reviewed them in 1965. Two years later some refinement was proposed for the difficult case of bone, to take account of the part of the bone in which deposition had occurred. No further significant changes were proposed in the 1969 ICRP review.

Despite the consistency of these ICRP recommendations concerning protection against the hazardous effects of radiation, much thought was going into obtaining a better understanding of the mechanisms of energy depositions. As a result, it was becoming clear that "dose" and "LET" were mean concepts, and account needed to be taken of the statistical fluctuations of the loss of kinetic energy of the charged particles as they passed through matter. These new ideas were crystallized in the 1971 Report of the ICRU [ICRU 19, 1971], which defined the new terms "specific energy" and "lineal energy." Thus, the absorbed dose, D, was redefined as a mean energy deposited in an elemental volume. Despite these radical changes, no fundamental alterations were proposed for applying them to radiation protection.

Two new concepts were introduced, however, viz. absorbed dose index, D_I, and dose equivalent index, H_I. These were defined as the maximum absorbed dose and the maximum dose equivalent, respectively, within a 30-centimeter (cm) diameter sphere centered on the point of interest. The sphere was to consist of material equivalent to soft tissue with a density of 1 gram/cm^3. Their purpose was to circumvent the difficulty that it was rarely possible to measure the dose (or dose equivalent) in an organ directly; this needed to be ascertained indirectly. Essentially, this could be done by measuring the exposure free in air and then making calculations using appropriate absorption coefficients.

In its 1977 recommendations [ICRP 26, 1977], the ICRP adopted SI units for radiation protection. Thus the unit of absorbed dose was to be the gray (1 Gy = 1 J/kg) and that of dose equivalent was to be the sievert (Sv). Because the quality factor Q (previously QF) was dimensionless, the sievert also had units of joules/kilogram (J/kg). The relationship between the value of Q and LET_∞ remained unchanged. However, completely without explanation, the ICRP raised the Q value for alpha particles to 20.

A major innovation was the recognition that body tissues vary in their sensitivity to the effects of radiation. The ICRP therefore introduced a tissue weighting factor, w_T, which represented the proportion of the stochastic risk resulting from a specific tissue (T) to the total risk. Values of w_T for six major regions of tissue were recommended; those not specified were lumped together as "Remainder" to which an additional value of w_T was assigned. The sum of all the separate values of w_T was made equal to 1.

The ICRP noted that the relationship between the dose of radiation received by an individual and any particular effect was a complex one. They had therefore decided to continue with the simplifying assumption that this relationship was linear with no threshold. With this assumption it was justifiable to consider dose as the mean dose over all the cells of uniform sensitivity in a particular tissue or organ. The use of the mean dose had practical advantages in that the significant volume could usually be taken as that of the tissue or organ under consideration.

The ICRP 1977 report also repeated its warning of 15 years earlier concerning the irradiation of nonhomogeneous tissue. It reminded readers that in such cases the use of mean dose over the tissue ceases to be strictly valid, and once again gave the example of radioactive particulates in the lung.

During the following decade, only comparatively minor amendments and extensions were published. In the 60 years since their first meetings the ICRP's and ICRU's recommendations had grown from a few common-sense observations to a highly formalistic set of documents. The recommendations published at the beginning of the 1990s have, however, proved to be the most controversial. These will be discussed in Chapter 3.

3.

The 1990 Recommendations of the ICRP and Their Consequences

The tacit agreement between the ICRP and the ICRU on radiation units and their application to radiological protection was rudely shattered in 1990 when the ICRP adopted a totally new set of recommendations [ICRP 60, 1990]. Although these recommendations were still linked to the idea of absorbed dose as a fundamental quantity, the derived quantities were changed both in nature and in name.

The ICRP continued with the simplifying assumption that the relationship between the dose and the probability of a subsequent effect was a linear one with no threshold. This was described as "a reasonable approximation over a limited range of dose" but, as many of the arguments in the field of radiation protection are caused by nonlinearities in the dose-response relationship, this assumption must surely be questioned.

Once again, the ICRP argued that "dose" should be used as the average taken over the tissue or organ as a whole. To distinguish this from the dose, D, that was specified at a point, the symbol "D_T" was introduced to denote the tissue dose. The problems arising from nonuniform distribution were also briefly alluded to in the same paragraph (page 5), but otherwise no mention was made of the well-known difficulties created by alpha emitters in lung or bone.

It was in the nature of the allowances to be made for different types of radiation that the ICRP's recommendations proved to be most controversial. Such allowances were to be made by introducing a new modifying factor, called the radiation-weighting factor, w_R. It was to be selected for the type and energy of the radiation *incident on the body* (our emphasis) or, in the case of a source within the body, emitted by the source. From these factors, another new quantity to be called "equivalent dose" was defined as

$$H_T = \sum_R w_R D_{TR} \tag{3.1}$$

where D_{TR} is the absorbed dose averaged over the tissue, T, due to radiation, R. The potential for confusion between this new term and "dose equivalent," used for many years, appears to have been disregarded.

In justifying the replacement of the quality factor Q by the radiation weighting factor w_R the ICRP explained that it believed that the detail and precision inherent in using a formal relationship between Q and the value of the linear energy transfer L to modify absorbed doses was not justified because of the uncertainties in the radiobiological information. Therefore, the ICRP had selected values of w_R based on a review of the biological information, a variety of exposure circumstances, and an inspection of the traditional calculations of the ambient dose equivalent. The strange implication that these had hitherto been neglected seems to have escaped the ICRP's notice. Although the values of w_R tabulated were no longer related to L but specified for particular types of radiation, the numbers concerned were effectively almost unchanged. Thus, for photons of all energies, electrons, and muons (i.e., "low-LET" radiations) w_R was set to 1. For alpha particles, fission fragments, etc., (i.e., "high-LET" radiations), w_R was set to 20. The only real innovation was the specification of w_R values for neutrons in various energy ranges.

Having introduced the new concept of w_R, the ICRP had to create another to allow for the effects of different tissue sensitivities. In place of the previous effective dose equivalent, it now specified a quantity designated "effective dose". This was defined as the equivalent dose weighted by the appropriate tissue weighting factors, w_T, i.e., a doubly weighted absorbed dose. Apart from giving numerical values to tissue and organs previously omitted, the values of w_T were effectively unchanged from 1977.

Another major change was in how to specify a limit in the risks to which an individual may be subjected. In its earlier 1977 recommendations [ICRP 26, 1977] for dose limits applied to occupational exposure, the ICRP used a comparison with the rates of accidental death in industries not associated with radiation. Recognizing that such comparisons were unsatisfactory for a number of reasons, in 1990 it again tried to establish a level of dose above which the consequences for the individual would be widely regarded as unacceptable. For this purpose the limiting dose could be expressed as a lifetime dose received uniformly over the working life, or as an annual dose received every year of work, without prejudice to the way in which the dose limit is finally specified. On the basis of the available data, the ICRP concluded that its dose limit should be set in such a way and at such a level that the total effective dose received in a full working life should not exceed 1 Sv received moderately uniformly year by year. It further recommended a limit on effective dose of 20 millisieverts (mSv) per year averaged over 5 years, with the further provision that the effective dose should not exceed 50 mSv in any single year. These recommendations represented a significant reduction to the previous limit of 50 mSv per year during the working life.

Many criticisms have been leveled at ICRP 60. First and foremost, it split the radiation community between those who felt obligated to adopt the new modifying factors w_R and those who wished to retain the traditional quality factors Q. Indeed the British Committee on Radiation Units and Measurements (BCRU) felt sufficiently strongly about the matter to issue separate advice early in 1993 [BCRU Memo, 1993], and it is worthwhile to repeat its comments in detail.

They noted that the operational quantities "ambient dose equivalent," "directional dose equivalent," and "personal dose equivalent" were all suitable for metrology, and for comparisons with quantities previously defined by the ICRP that indicate the risk resulting from exposure to ionizing radiation. The ICRP had, as noted previously [ICRP 26, 1977], recommended that the limiting quantities be "effective dose equivalent" and "organ dose equivalent"; subsequently, it had been shown that these ICRU measurable quantities could be related to the ICRP limiting quantities. The question implicit in these observations, "Why change?", was not directly asked.

Their next observation concerned the factor w_R. This was *not* defined in terms of the radiation field in the organ in question. From the point of view of metrology and calculation, this gave rise to ambiguity because no allowance was made for the change in the spectrum between the radiation entering the body and the radiation reaching the organ in question. In addition to such ambiguity, the new ICRP quantities (equivalent dose and effective dose) did not have the characteristics recommended by ICRU, viz. that they should be additive, point specific, and routinely measurable.

Similar criticisms of the 1990 recommendations were made by members of the Italian radiation physics community [Pelliccioni and Silari, 1993]. They were partly concerned with the proposed introduction of yet another new set of quantities (perpetuating the chronic instability of the radiation protection dosimetry systems and giving rise to confusion) and partly with the fact that the quantities defined for radiation protection were not "measurable quantities" as usually defined in physics.

The ambiguity in the definition of the term "average organ dose" was also commented on. This was defined as the ratio of the total energy imparted to the organ or tissue and the mass of the organ or tissue. It is not clear whether this should be read as "total-energy imparted" or "total energy-imparted". In the former case the term is imprecise, if not incorrect; in the latter it is definitely wrong because the "energy-imparted" is a stochastic quantity and therefore has to be associated with a distribution of values rather than an indefinable *total* value. Fortunately, the ICRU resolved this ambiguity soon afterwards.

The writers further returned to the point that the radiation weighting factors, and the quantities such as the organ equivalent dose defined through them, are unsuitable in practice. This is because, in a measurement, the contributions from the different components of the radiation field cannot be distinguished. In a multicomponent radiation field (as is often the case with neutrons) the equivalent dose to a given organ does not coincide with the sum of the values of the individual equivalent doses delivered to the organ by the various components of the field.

Finally, the choice of values of w_R was criticized. The maximum value for neutrons, 20, was both lower than the value given by ICRU, which was 25, and also much lower than values suggested by some radiobiological investigations (and much higher than Lea's data). Some of these investigations indicated that values of RBE for certain effects may approach 200 at very low doses, although it has also been argued that, in such cases, the concept of "dose" (and hence of RBE) is unsound anyway. This point will be returned to later in this chapter.

Many comments of a similar nature have been made privately by eminent radiation scientists to the present authors. The majority of these comments concern w_R, both in principle and in detail. In principle, w_R applies to the radiation incident on the body, but this is to be used to multiply the mean dose to the organ, D_R. Clearly, the multiplication of the dose in one location by a factor that specifies radiation quality at another location is not only biologically unreasonable but also makes the equivalent dose virtually unmeasurable. In detail, the way in which the values of w_R are assigned to different energies of neutrons implies that the neutron spectrum must be known to a much greater degree of accuracy than can be expected in practice. Hence, the person involved in making measurements and calculations for radiological protection is in a worse position than before.

In an attempt to clarify the situation, a Joint Task Group of ICRP and ICRU members was subsequently set up to address some of these criticisms. The ICRP report [ICRP 74] was published in 1996 and provided a compilation of conversion coefficients to link the physical, operational, and limiting quantities as defined by the two commissions. (The corresponding ICRU report [ICRU 57] has still not been published at the time of writing this book in 1998.) Having given advice on the recommendations of ICRP 60 in 1993, the BCRU updated its advice in 1997 [BCRU Memo, 1997] following the publication of ICRP 74 in 1996.

Essentially, the two sets of BCRU advice were similar. However, the 1997 update did consider the case of neutron dosimetry in more detail and noted the wide divergencies that could occur between the effective dose, E, and the operational quantity personal dose equivalent, $H_p(d)$. (The latter is defined as the dose equivalent in soft tissue, defined as in the ICRU sphere below a specified point on the body at depth, d, that is appropriate for strongly penetrating radiation.) These divergencies vary with the energy of the incident neutrons, and it was therefore recommended that it should be specifically noted, with the dose record, when narrow-band, intermediate energy neutrons and neutrons of energy >50 million electron volts (MeV) have been monitored. The BCRU also alluded to the fact that it had made no headway with its concerns about the impact of the dose limiting quantities in terms of radiation weighting factors rather than quality factors proposed by ICRP 60.

Let us review the steps leading to the present set of correction factors. First came the relative biological effectiveness, initially specified in terms of average specific ionization and then in terms of linear energy transfer. Because of the necessity of separating biological experimental results from recommendations for radiological protection, the RBE was replaced by a set of quality factors Q, values of which were assigned by a committee to ranges of LET. Now Q is replaced by w_R to which numerical values are assigned in terms of particles and their energies. The parallels between these changes and the development of correction factors required by early astronomers is striking; eccentricity, deferents, and epicycles were postulated to account for observations based on the false premise of a geocentric universe, and an increasingly complex series of corrections (weighting factors) have had to be introduced in vain attempts to support an unsound premise.

In the present context, the "unsound premise" is that absorbed dose is a fundamental concept that can be used as an effective predictor of radiation effects. In some particular cases this might be true, but in many cases this is not so. It is as if one were to suggest that the relationship between force and acceleration can change its form according to the nature of the force and the nature of the matter being accelerated when, in fact, it is always linear. In our astronomy analogy, it would not be too difficult to predict the motion of the Sun and the planets around the Earth "to a reasonable degree of approximation" if one were to use different values of the gravitational constant for each of the heavenly bodies observed.

Criticisms of the use of absorbed dose as a basis for assessing the effects of low levels of radiation are not new. At the 17th meeting of the NCRP in 1981, V. Bond, the Head of the Medical Department of the Brookhaven National Laboratory, observed that for stochastic processes such as the induction of cancer at low levels of radiation, it is the effect within a cell (or a small number of cells) that is important. However, because at low levels of radiation (i.e., those of significance in radiation protection) a large proportion of the cells will have received no radiation, the mean dose per cell represented by the average tissue dose is not the same as the mean dose per dosed cell. A better quantity to use in this context is the fluence of charged particles through the critical volumes. Only when all the cells have received at least one hit (i.e., at "doses" of ~10 cGy for low-LET radiation and ~1 Gy for high-LET radiation) does dose become a suitable surrogate for charged-particle fluence.

The incidence of cells hit, I_H, for a given exposure can be written as

$$I_H = \Phi \, \sigma \qquad (3.2)$$

where Φ is the fluence of charged particles and σ is the cross section of the critical volume of the cell. The following argument, based on that developed by Bond and his various co-workers during the 1980s, will show that Φ and not the absorbed dose, D, is the conceptually appropriate quantity for the amount of low-level radiation (LLR) involved in an exposure.

Let $\bar{z}$ be the mean specific energy, due to single hits only, for the fractional number of relevant cells hit. Since $\bar{z}$ is an average over only the critical volumes of the relevant cells hit while D is conceptually an average over all cells in the sample,

$$D = \frac{z_1 + z_2 + \cdots z_H}{N_E} = \frac{z_1 + z_2 + \cdots z_H}{N_H} \cdot \frac{N_H}{N_E} = \bar{z} \cdot \frac{N_H}{N_E} \qquad (3.3)$$

where N_H and N_E are the numbers of hit and exposed cells, respectively, as indicated by microdosimetric measurements. With large amounts of radiation, all exposed cells are hit; i.e., $N_H = N_E$ and thus $D = \bar{z}$. With LLR, however, $\bar{z}$ remains constant with increasing Φ or D and only the proportion of cells hit increases with increasing Φ.

Experimental verification of these concepts was subsequently obtained for lung tissue irradiated with alpha particles [Simmons and Richards, 1984, 1989]. Using material derived from rat, dog, and human lung, Simmons and Richards were able to follow the energy deposition patterns as the alpha particles traversed the tissue. From these patterns, distributions were obtained for (amongst other things) the volume of tissue receiving a particular specific energy. Initially, the volume of tissue was created in a mathematically simple manner that bore no relation to any biological feature; subsequently, volumes (analogous to the "critical volumes" of Bond and his co-workers) were taken corresponding to those of the whole cell and of the nucleus.

The relevant results for a sample of human lung are shown in Figure 3.1(a-c). At the lowest fluence of alpha particles, only a very small volume of cells and an even tinier volume of nuclei are hit. Those targets that are hit receive higher specific energies than would be expected from the distribution to the tissue overall.

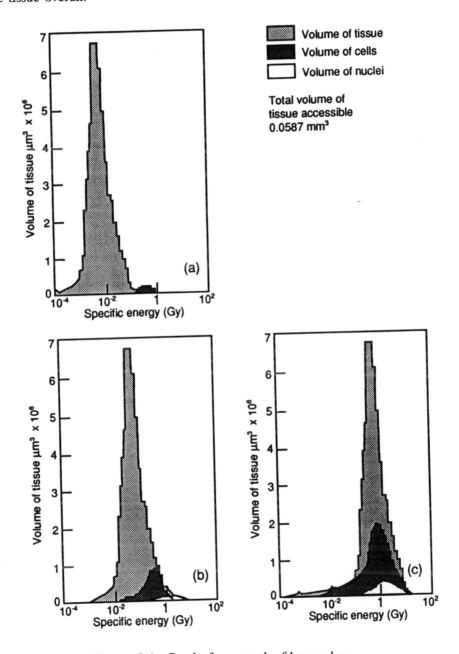

Figure 3.1. *Results for a sample of human lung.*

With an increased fluence, an increased volume of cells and nuclei are hit. Most targets are still receiving only one hit, so the positions of the peaks and the shapes of the distributions do not significantly change from those seen at the lower fluence. Because the target size for nuclei is smaller than for cells, the distribution of specific energy is centered at a higher value for the former compared with the latter.

At the highest fluence, most cells receive more than one hit. This means that there is less variability in the specific energy deposited in each cell because it is the sum of several events. As a result the curve shifts to higher specific energies and begins to form a narrower distribution. Furthermore, the peak in the curve for cells now approximates the curve for tissue, showing that target size no longer totally dominates the specific energy distribution. However, except in the rat sites (not shown) the distributions for nuclei still do not line up with those for cells; this situation is not reached until the fluence of alpha particles is increased by approximately another factor of 3, and corresponds to the point where $D = \bar{z}$. This occurs at a value of D of approximately 2 Gy, which, possibly fortuitously, is in fair agreement with the value predicted by Bond and Varma [1982].

The results obtained by Simmons and Richards [1989] clearly confirm the view propounded by Bond and Varma [1982] that with LLR, absorbed dose is conceptually not a "dose" to the cell's critical volume of interest. Rather, it more closely represents the average energy concentration in the medium from the total amount of energy carried by all of the charged particles that pass in the vicinity of the cells. Only a fraction of this total amount of energy interacts with and is transferred to the critical volumes. Any system of measurements for radiation protection should recognize this fact.

The specific energy may, of course, be due to one or more energy deposition events. The distribution function of the specific energy deposited in a single event, $F_I(z)$, is the probability that a specific energy less than or equal to z is deposited if one event has occurred. The probability density, $f_1(z)$, is the derivative of $f_1(z)$ with respect to z:

$$f_1(z) = \frac{d\, F_1(z)}{d\, z} \; . \tag{3.4}$$

(It is also known as the single event distribution of z.)

The expectation value

$$\bar{z}_F = \int_0^\infty z\, f_1(z)\, d\, z \tag{3.5}$$

is called the *frequency mean specific energy per event.* In the ICRU Report 36 on microdosimetry in 1983 it was suggested that the term "low dose" can be used if the absorbed dose is < 20% of $\bar{z}_1$. A simpler and virtually equal specification is to define a dose as "low" if the fraction of targets affected by the radiation is < 20%.

Consider the implications of this definition of low dose for the case discussed previously of inhaled alpha-emitter irradiating lung cells. If one makes the simplifying assumption that the nuclei of such cells can be represented by spheres of diameter 7 μm, and that the LET of the particles is 100 keV/μm, then using the formula $\bar{z}_F = 0.20 \cdot L/d^2$ [Appendix A of ICRU 36] gives $\bar{z}_F \cong 0.4$ Gy. If "low" implies a value of dose < 20% of this, the result is an approximate limiting value of D = 0.08 Gy. Attempts to produce an RBE for alpha particles at doses below 0.08 Gy are really trying to compare the outcomes of two different and noncomparable irradiation conditions: for the alpha-irradiated cells, < 20% of them are actually being irradiated; for the cells undergoing low-LET irradiation, all will receive at least one hit at 0.08 Gy. At doses below ~0.01 Gy this is no longer true even for low-LET radiation, so comparisons become even more meaningless. One must consider the effects of high-LET (charged particle or neutron) irradiation in absolute, not relative terms.

As a result of further work with his colleagues, Bond was able to show another fundamental weakness of dose in two 1991 publications a decade later [Bond, 1991; Bond et al., 1991]. Based on his calculations using the revised dosimetry of the atomic bomb survivors, Bond concluded that *total* energy (rather than energy per unit mass) was a more realistic parameter to use in a radiation protection measurement system. Bond insisted that attempts to relate "risk" to the dose to an individual were inappropriate, as risk could only be measured as a property of a population. Using a plot of attributable cancers other than leukemia against total energy, it was shown that a minimum value of "collective energy" existed, below which no cancers were found. The value of this minimum energy (which was not necessarily a threshold as Bond et al. were careful to point out) was about 3.5 kJ. If one then tries to reverse this type of calculation, i.e., to convert back to the dose for a standard 70-kg man, one arrives at a figure that is about 10 times the mean lethal dose. Clearly, the concept of dose is not an appropriate one from the point of view of radiation protection. Possible alternatives are considered in the next chapter.

4.

Simulation Models of Biophysical Effectiveness

A critique of the system of dosimetry currently used for radiation protection has been made in the preceding chapters. If this has any substance, then deficiencies in the proposed mechanisms of induced damage should be manifest at the most fundamental level, particularly in the role of the basic quantities of absorbed dose, LET, and RBE. An indicator that there may indeed be serious deficiencies is the overabundance of models of cellular damage. All the models are based on different concepts; although all can be supported by some experimental evidence, none have achieved an adequate degree of success. Chapter 4 broaches the possibility here that the deficiencies are associated with the capability of the various track structure parameters that serve as descriptors of the initial physical damage in biomaterial. The following sections give a brief survey of the principles and concepts used in some of the radiation damage models, then explore at depth the nature of the biophysical mechanisms of radiation action. Biochemical models specifically concerned with repair and protective mechanisms are excluded.

Some Models of Bioeffectiveness

Development of mathematical models of radiation action is of continuing interest, particularly in the important role of establishing formally the key links between exposure and effects. Objectives of modeling are multi-fold. Models can help to identify and interpret damage mechanisms. They can be used to identify physical quantities that, ideally, should enable correlation of biological data into a unified scheme for different radiation types and biological endpoints.

Importantly, for practical implementation of a system of protection, they can give guidance on the desired response functions for the instrumentation needed for the practical measurement of bioeffectiveness. Any serious model should have predictive properties and be applicable to the extrapolation of biological effects observed at higher doses as well as the much lower natural environmental doses at which biological experiments and epidemiological studies are not yet statistically significant. An adequate model should be able to interpret and explain complex radiation effects; for example, the observed inverse dose-rate effects [Hill et al., 1982; Sykes and Watt, 1989]; the action of Auger electron emitting radionuclides incorporated into mammalian cells as used in nuclear medicine [Humm et al., 1994]; the effects of the duration of exposure and dose-rate effects; and repair of damage [Hall, 1988]. It should be able to cope with both direct physical interactions and the indirect chemical action known to be caused by diffusing radicals [Chadwick and Leenhouts, 1981]. A model of bioeffectiveness should also have the capability to assist optimization of quality in conventional electron and photon therapy and in the newer, high-LET therapies based on boron neutron capture, accelerated heavy ions, and fast neutron irradiations [Wambersie, 1990].

Models of biological damage of interest for radiation protection tend to be of two categories: cellular models and models of carcinogenesis. Only a selected number of the more fundamental types of cellular model are surveyed briefly here to illustrate the basic concepts and mathematical formulations used and to serve as background information prior to presenting arguments that energy deposition cannot be relevant to the quantification of biological damage mechanisms. Other critical reviews and descriptions of some selected models are available in the literature (see *Selected Readings* in the *References* section).

Hit and Target Theory

Hit and target theory is of historical importance because it is a rudimentary component of virtually all subsequent radiation damage modeling. Its conceptual success lies in the use of a simple Poissonian statistical treatment to account for the pure exponential and sigmoid (shouldered) shapes of the observed dose-response curves that are attributable to the discrete random interactions along the radiation tracks and to the natural distribution in size of the individual biological cells. Zimmer [1961] discusses the concepts of hit and target in detail. A "hit" is largely undefined. It can be the number of ionizations or the amount of energy deposited per unit volume. The volume, or mass, is that of a radiosensitive target. For ionizing radiations the definition of the "hit event" is taken according to context and variously means "an ionization," quantified by "ionization per unit volume" or "formation of primary ionizations, meaning ion clusters along the charged particle tracks, etc." For practical expediency, because of the difficulty in measuring ionizations in a biological entity, the hits are usually expressed in terms of the corresponding energy deposition, usually in electron volts (eV), which has a convenient parallel with the chemists' G value (yield of species produced per 100 eV). "Targets" usually refer to the net volume, or mass, of the radiosensitive region. For enzymes and some viruses the region may be the whole volume of the structure or biomolecule. For mammalian cells the volume will be equivalent to that of the sum of the radiosensitive regions within the cell nucleus. The "target volume," sometimes termed the "critical volume," may differ from the true volume of the biological entity; it can be derived from the observed semilogarithmic dose-response curves, provided that allowance can be made for known modifying factors such as indirect chemical action, dose-rate effects, oxygen sensitization, and repair. The "dose" can be expressed as "hits per unit volume or mass" or "passage of particles per unit area". In the case of passage of a fast charged particle, a formal action cross section, S, can be defined as

$$S = q\,(1 - e^{-h}) \tag{4.1}$$

where q is the mean geometrical cross-sectional area of the structure containing the radiosensitive target and e^{-h} is the probability that the ionizing particle can traverse the true target without interaction when the mean number of hits is h [Dertinger and Jung, 1970].

The General Equation for Hit and Target Theory

The general formula represents the case in which the exact number of n hits in each of m targets is required to produce the lesion [e.g., Zimmer, 1961]. If the expectancy value for the number of stochastically fluctuating hits is h, then the biological surviving fraction, $F_{n,m}$, for n hits in each of m targets is

$$F_{n,m} = 1 - (1 - F_{n,1})^m \tag{4.2}$$

where

$$F_{n,1} = e^{-h} \cdot \sum_{r=0}^{n-1} \frac{h^r}{r!} \tag{4.3}$$

is the Poisson probability of a single target surviving up to n hits.

The cases of particular interest are single hit, single target ($n = 1$, $m = 1$) inactivation and single hit, multi-target ($n = 1$, $m = m$) inactivation.

Single Hit, Single Target (n = 1, m =1) Inactivation

The survival fraction, $F_{1,1}$, of biological structures inactivated by a single hit in a single target is:

$$F_{1,1} = e^{-h} \tag{4.4}$$

for which the hit-response curve is characterized by a straight line of gradient -1. When $F_{1,1} = e^{-1}$, the fractional survival is 0.3679 (37%) leading to the well-known fact that there is, on average, one hit per target at the D_{37} dose in a dose-response curve. Here the surviving fractions are expressed as functions of the number of hits, to preserve generality, but in practice they would be given as functions of absorbed dose, D, or particle fluence, F. Simple relationships connect these quantities. Thus, $h = D/D_{37} = k \cdot D$ where k is the radiosensitivity and $h = \sigma \cdot \Phi$ with the action cross section for production of the lesion, $\sigma = 1/\Phi_{37}$. Φ_{37} is the particle fluence for 37% survival.

Single Hit, Multitarget (n = 1, m = m) Inactivation

If in a multiplicity of m targets each of which requires a single hit to induce the effect in the biological entities under test, then the fraction surviving, $F_{1, m}$, is

$$F_{1,m} = 1 - (1 - e^{-h})^m . \tag{4.5}$$

For large average target numbers, h, the above equation can be expanded using the binomial theorem to produce a good approximation to the yield:

$$F_{1,m} = m \cdot e^{-h}$$
or
$$\ln (F_{1,m}) = \ln (m) - h \tag{4.6}$$

which gives hit-response curves that are sigmoid in shape with a constant gradient of -1 at large hit numbers. The backward extrapolation of the gradient intercepts the ordinate at the value of m, the target multiplicity. Idealistic examples of single hit, multitarget response curves are shown in Figure 4.1.

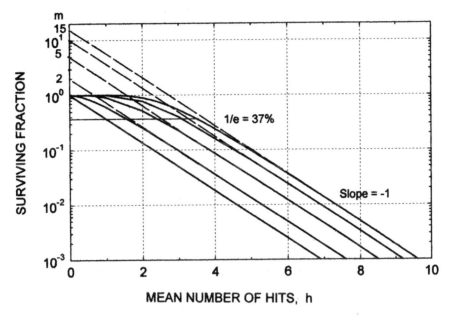

Figure 4.1. *Hit-response curves are shown for target multiplicities 1, 2, 5, 10, and 15.*

Basic target theory successfully incorporates statistical concepts to reproduce the general shapes of high-LET (typically linear dose-response) and low-LET (typically sigmoid dose-response) survival curves and continues to influence damage modeling. However, in their basic form a number of predictions are not adequately supported by observation. These are, for example, the predictions that the sigmoid dose-response curves for multitarget action will have zero slope at low doses; have a constant linear slope at high doses; and different radiation types with the same LET or ionization density will produce identical effects. The action cross section for production of the biological effect (equation 4.1) for single target, single hit action is expected to "saturate" at the geometric cross section, but it is well known that there are many occurrences where the cross section may be more than an order of magnitude greater than the geometric. The excess is due to the action of delta rays around fast heavy particle tracks for single hit, single target kinetics in, for example, enzymes and some viruses [Dertinger and Jung, 1970]. In the case of mammalian cells the cross section excess is not observed until the heavy particles slow down to near the end of their ranges but where they still have sufficient energy to generate an intense yield of energetic delta rays (see *The Role of Delta Rays Associated with Fast Heavy Ion Tracks* in Chapter 5).

Lea [1955] considers that target theory is applicable to those biological effects caused by the production of ionization by radiation in, or in the immediate vicinity of, some particular molecule or substructure; i.e., the targets. Typical biological endpoints to which the target theory may be applied are, for example, inactivation of viruses, gene mutation due to ionization of the gene molecule or chromosome breakage following transit of a particle track through the chromosome. The theory is not appropriate in situations where the biological damage is not localized as, for example, in a toxic chemical environment. In addition, caution should be exercised in application of target theory and it should not be used to interpret effects *solely* on the basis of survival curves. Lea suggested that three criteria be used to help avoid erroneous interpretation: (1) determine the manner in which the number of organisms or cells affected increases with dose; i.e., the dose-response curve; (2) determine the manner in which the effect is dependent on dose-rate; (3) investigate the RBE dependence on radiation type and energies. For RBE dependence it was usually considered sufficient to study radiations up to and including alpha particles; but, as will be demonstrated later, there is often an approximate proportionality relationship among commonly used quality parameters. Thus it is essential to include accelerated heavy ions up to argon and greater mass numbers. (RBEs are discussed in more detail in Chapter 5. See especially Figure 5.8.)

Lea's Associated Volume Method in Target Theory

Lea [1940] was among the first to apply details of the charged particle track structure, within the ambit of target theory, in an attempt to interpret and explain biological damage mechanisms. Physical properties of heavy particle track structure thought to play a role were the mean free path between clusters of primary ionization produced along the particle track, the yield and spatial distribution of the associated delta rays, and the possibility of the particle track producing radicals by interaction with bound water inside and immediately outside the chromosomes, thereby enabling biological damage to be induced at a distance determined by the diffusion length. Enzymatic repair of damage was recognized as an important modifying factor. The hit was defined as "production of ionization in the target." Secondary ionizations associated with the delta rays emitted along the main particle track were treated separately from the primary ionization. Saturation damage, attributed to the occurrence of excess ionizations above the single ionization assumed necessary to cause the damage to the target, was accounted for in an "overlap factor" that was estimated by determination of each ionization's associated volume. The aim of the associated volume method is to determine the target size in the expectation that it could be associated with a significant biological structure and thereby prove presumed mechanisms of damage. The mean number of ion clusters produced per chord of length $2x$ through the spherical target is $2x/L$. Allowing for the randomness, the probability that one cluster will be produced is $(1-\exp(-2x/L))$ and the mean number of hits, N_h, per target of diameter d is given by:

$$N_h = 2\pi n L \int_{0}^{d/2} (1 - e^{-2x/L}) x \cdot dx \qquad (4.7)$$

where L is the mean free path between specific primary ionization clusters and n is the number of ion clusters per unit volume. Putting $\zeta = d/L$, and integrating gives:

$$N_h = \frac{\pi d^2 \cdot n \cdot L}{4} \left\{ 1 - 2\left(1 - e^{-\zeta}\right)/\zeta^2 + 2e^{-\zeta}/\zeta \right\} = \frac{\pi d^3 n}{6F} \qquad (4.8)$$

where $\qquad F = \left(\frac{2\zeta}{3}\right) / \left\{ 1 - 2\left(1 - e^{-\zeta}\right)/\zeta^2 + 2e^{-\zeta}/\zeta \right\}$.

The "dose" delivered to a sphere of diameter d is $n \cdot \pi d^3/6$ ion clusters, and because the mean chord distance through a sphere is $2d/3$, the average number of ion clusters produced along a charged particle track is $2d/3L$, assuming the particle has sufficient range. The mean number of hits produced for a dose averaging one ionizing particle per target is simply $2\zeta/3F$ as can be seen from equation 4.8 above. For a high-LET particle L is $<< d$ and the mean number of hits, N_h, tends to 1. For an average density (dose) of one ion cluster per target volume of $\pi d^3/6$, $1/F$ represents the mean number of hits per target.

Calculation of the associated volume for electrons and delta rays involves the following procedure. Electrons less than 100 eV are treated as single clusters equivalent to one ionization with a surrounding spherical volume of diameter d no larger than that for a single primary ionization. Electrons greater than 100 eV are treated separately as delta-ray tracks. Spheres are drawn around each ionization in the delta-ray track (omitting the primary ionization at the point of origin), then applying the basic formula, $n\pi d^3/6F$ (F is the overlap factor, equation 4.8) to determine the associated volume. The net result is added to the associated volume already determined for the primary ionizations in the primary track. A numerical integration technique is used to calculate the associated volume for any energy of delta electron. Lea has tabulated data for delta-ray energies from 0.1 to 480 keV and for target sizes in the range of 4 to 80 nanometers (nm) to 80 nm to cover the anticipated dimensions. Consistent estimates of target size are obtained for small viruses, enzymes, etc., but not for mammalian cells. The reader is referred to Lea's book [1955] for arguments supporting the conclusions that: good target sizes can be accurately determined only for small viruses and some bacteriophage but not mammalian cells; that high-LET radiation rather than low-LET radiation is more likely to give correct results; and that the assumption on target sphericity is unlikely to cause significant error. Subsequent interest in the associated volume method in application to mammalian cell lines has been limited. Possible reasons for the limitations are that although it is natural to treat each ionization identically in a medium assumed to be homogeneous, no allowance is made for the capability of different radiation types to deliver that ionization across intracellular interfaces. Also, on what could be an important point of detail, the values of the primary ionization per micron of tissue (i.e., a *volume* quantity sometimes called the ion density or specific ionization, which includes ionization from the delta rays) listed in Tables 11, 12, and 13 of the second edition of Lea's book [1955] differ by factors of 2 to 20 times from the different quantity, the *linear* primary ionization, which is equal to the yield of delta rays per unit track but specifically excludes their spatial contribution to the ionization (see Chapter 5 in this book). The difference between the linear primary ionization and the specific primary ionization density is an important one in the present context and the two should not be confused [Perris et al., 1986; Belli et al., 1989]. Lea's primary ionization is proportional to LET for a constant value of W, the mean energy required to produce an ion pair. z^2/β^2 is proportional to the linear primary ionization, λ, but applies only to fast particles, as discussed in *Katz' Model for Cellular Inactivation by Heavy Ions* later in this chapter. See also equation 4.23.

Lea's Theory for Yields of Chromosome Aberrations

A major application of target theory is in the study of radiation damage to chromosomes, producing chromosome exchange aberrations, because of their potential influence on genetic abnormalities and, indeed, the possibility that they were precursors to lethality in mammalian cells. Lea reasoned that passage of a densely ionizing particle such as a proton or low-energy electron anywhere through the chromatid causes a break; i.e., in ~100-nm dimension, and that to cause this break some form of energy transfer had to be involved. One mode of energy transfer, attributed to Gray in 1940, could be via the diffusion of water radicals. The ionization produced in the water bound inside and immediately outside the chromosome could produce active radicals that might diffuse to interact and rupture chemical bonds. If this is so, then effects due to water radical interactions may occur at diffusion distances up to 15 nm from the particle path. On the other hand, current thinking is that hydroxyl radicals (OH), believed to be the most important, have diffusion distances of only 2 to 3 nm, because of the presence of fierce scavenging, thereby effectively localizing the action to the immediate vicinity of the target chromosome [Chatterjee and Holley, 1991].

After careful consideration of the available experimental information, Lea developed a model for the induction of chromosome aberrations in eukaryotic cells by X-irradiation. For short irradiations, all primary breaks in chromatids are assumed to coexist in the cell nucleus, some of which will interact to produce exchanges. For prolonged irradiations, delivered either continuously at low dose rates or in fractions, many of the breaks may be repaired before sufficient time has elapsed for other breaks to be produced with which the earlier breaks may interact and create exchanges in the vicinity. The yield of exchanges should therefore decrease with the duration of the irradiation. To model the process it is necessary to allocate a mean time, τ, for repair of breaks. The calculation then proceeds in a manner analogous to bimolecular reactions in chemistry. Thus, the rate of change of primary breaks with time, dn/dt, at any arbitrary time, t, is equal to the production rate of breaks, which is proportional to dose rate (ζI) minus the mean rate of loss of breaks due to repair (n/τ). The corresponding differential equation and its solution for the number of breaks n at any time t are

$$\frac{dn}{dt} = \zeta \cdot I - \frac{n}{\tau}$$

$$\text{whence,} \quad n = \zeta \, I\tau (1 - e^{-t/\tau}) \quad .$$

(4.9)

If the duration of the irradiation is T, then at any time t after the end of the irradiation, $t > T$, the number of breaks will decrease as

$$n = \zeta \, I\tau (1 - e^{-T/\tau}) \cdot e^{-(t-T)/\tau} \quad . \tag{4.10}$$

As exchanges require the interaction of two primary breaks, the yield, Y, of exchanges is proportional to n^2; i.e.,

$$Y = k \int_0^T n^2 dt$$

$$= \frac{1}{2} \zeta^2 k \tau (IT)^2 \cdot G \qquad (4.11)$$

$$\text{with} \quad G = 2 \left(\frac{\tau}{T} \right)^2 \cdot \left\{ \frac{T}{\tau} - 1 + e^{-T/\tau} \right\}$$

where k is a constant of proportionality. For a dose $D = I \cdot T$, the number of exchanges is proportional to $D^2 \cdot G$.

Lea obtained good agreement with experimental data on X-ray-induced chromosome damage in the sperm of the Drosophila fruit fly. A requirement of the theory to test the D^2 dependence is that the dose be varied by varying the dose rate for constant T and hence constant G. Furthermore, if the dose is kept constant and the dose rate varied, the yield of exchanges should be proportional to G. For the latter case, a plot of the number of exchanges against the duration of the irradiation should be identical in shape with a similar plot of G and will provide a simple method of determining the mean repair time, τ. In studies of repair, it is probably very important to conduct twofold experiments, one set keeping the duration of irradiation constant for varying dose rate and the other set keeping the dose rate constant, while varying the duration, T, of the irradiation if the subtleties of the repair action are to be properly assessed. This requirement seems to have been largely ignored in recent years.

Lea concluded on the basis of a comparison with X-rays and neutrons that a single particle could cause a chromatid break but more than a single ionization was necessary. (For reasons to be given later in Chapter 5, it should be remembered at this stage that the details of the DNA duplex had yet to be discovered. Therefore, a double-strand break (dsb) in the DNA could not therefore be considered a contending lesion.)

Theory of Pairwise Lesion Interaction (PLI) for Chromosome Aberrations

Explanation of the observed yields of chromosome aberrations in human lymphocytes and the trend with the temporal distribution of dose is based on the previous postulate that damage is due to a pairwise molecular reaction between two primary lesions in chromatin, in kinetic competition with enzymatic repair of these lesions [Harder and Virsik-Peuckert, 1984; Harder et al., 1991]. The pairwise interaction range of two molecular lesions in the chromatin was estimated by Lea to be 1 micron ± 10%. Also, this dimension required for the pairwise lesion interaction (PLI) is thought to be able to predict the radiation quality or track-structure effects due to the balance between the range of the interaction forces and the track-structure-dependent distances between radiation-induced lesions. Harder [1986] reinforces the claim that Lea's model can explain the temporal dose distribution and radiation quality effect on reproductive cell death, if one assumes that cell lethality is caused by PLI similar to or related to the exchange-type chromosome aberrations. This is called the "sublethal damage" pathway. It may coexist with the "potentially lethal damage" (may remain unrepaired) pathway. Harder advances the form of Lea's model in two important ways. First, he applies a more rigorous mathematical formalism that takes into account the stochastical distributions of energy deposition events using the principles of structural microdosimetry [see Rossi and Zaider, 1995]; second, he shows that the relative variance of the ionization density, as used by Lea, is directly proportional to the dose-restricted LET, $L_{D, 100}$, which in turn is used as a specifier of the radiation quality [Harder et al., 1988]. The empirical basis for Harder's PLI formulation is that the intratrack lesion interaction is in the nanometer region and that the intertrack interaction has a larger interaction range of dimensions less than that of the cell nucleus. PLI is assumed to occur in small "contact regions" between different chromatin fibers or between sections of the same fiber and these may be of temporary existence due to typical molecular movement. The intratrack effects are proportional to dose and the intertrack action is proportional to D^2, resulting in a yield dependence of the chromosome aberrations of the dual action form $\alpha D + \beta D^2$. From consideration of the probability per unit time of interaction of pairs of lesions, still present in an unrepaired state and therefore able to interact, the mean interaction rate, $\varepsilon(t)$, at any time, t, for each contact region is deduced to be

$$\varepsilon(t) = \frac{ak}{2}\left[m\left(\bar{n}_1^2 - \bar{n}_1\right) + m^2 \cdot \bar{n}_1^2 \right] \; . \qquad (4.12)$$

a is a proportionality factor and k represents the interaction efficiency of the contact between lesions. m is the mean value of the number of particle traversals, assuming they have a Poisson distribution of randomness. Equation 4.12 has the expected linear quadratic response. Passage of a charged particle through the targets will produce a stochastically fluctuating number of lesions, n_1 of which will be reactive at time t. To allow for the repair of the radiation-induced lesions, the mean value and moments of the n_1, reactive lesions, are taken as functions of the time interval, $t - \tau$, that has elapsed since the particle traversal at time τ. By introducing the probability function $p(t - \tau)$ that an ionization produced at time τ and resulting in a lesion at time t, and if i is the number of ionizations per particle traversal, then the distribution moments of n_1 are directly relatable to those for i. Also, because $m \cdot i$ is proportional to dose, $(c \cdot D)$, where c is the proportionality constant, equation 4.12 can be rewritten in terms of the ionization as moments of the ionization density thus:

$$\varepsilon(t) = \frac{akp^2(t-\tau)}{2}\left[\left(\frac{\bar{i^2}}{\bar{i}}-1\right)cD + c^2D^2\right] \quad . \tag{4.13}$$

The general form of the equation for the mean reaction rate is obtained by considering that the distribution of the n lesions present at time t, being the convolution of the lesion number distributions manifest at past elements of irradiation time $d\tau$ each combined with its own restitution factor, $p(t - \tau)$. The final generalized result for the lesion interaction rate, which is valid for continuous or fractionated irradiation, is

$$\varepsilon(t) = \frac{a \cdot k}{2}\left[\left(\frac{\bar{i^2}}{\bar{i}}-1\right)\int_0^\tau p^2(t-\tau)\,cD_r(\tau)d\tau + \left(\int_0^\tau p(t-\tau)cD_r(\tau)d\tau\right)^2\right] \quad . \tag{4.14}$$

D_r is dose rate.

Tests of the model are limited to the choice of data for radiations of atomic number $Z < 3$; i.e., up to alpha particles. Results are not yet available for heavier ions and are awaited with interest because of their greater discriminating ability in the choice of quality parameter. A plot of the relative variance in ionization density, $(\bar{i^2}/i - 1)$, against $L_{100,D}$ shows a linear relationship, indicating a direct proportionality between the linear primary ionization per unit track and $L_{100,D}$. The proportionality occurs because of the approximate invariance of the ionization yield per delta ray in the spatial region determined by the 100-eV cutoff. Thus, either the restricted primary ionization density or the dose-restricted LET with 100-eV cutoff appears to be a good quality parameter. Also, good agreement is obtained with the experimental results for radiation-induced exchange type chromosome aberrations and reproductive cell death [Harder et al., 1991] for radiations up to alpha particles. However, maxima RBEs are observed at different values of $L_{100,D}$ according to radiation type (see Figure 5.9), indicating that $L_{100,D}$ is not in fact a unifying quality parameter that would defeat the universality of the model.

Two-Component Models of Cellular Damage

For heavy charged-particle irradiations, two-component models satisfy the requirements for the possibility of a nonzero initial slope of the dose-response curve and the apparently important cumulative action of the delta-ray electrons to produce the biological lesion. The basic tenets of the hit and target theories are retained, but the radiation action is divided into two components: a high-LET component, representative of a semi-empirically defined track core, and possessing pure exponential survival characteristics ($F_{1,1}$, equation 4.4), which is typical of noncumulative, irreparable damage; and a low-LET component, which acts cumulatively on multiple targets to produce sigmoid-type survival ($F_{1,m}$, equation 4.5). The low-LET component or sigmoid-type survival is attributable to the delta-ray penumbra surrounding the ion track. This damage is considered to be repairable, but explicit repair mechanisms are not included in the two-component models. Many independent attempts have been made to develop models of cellular repair mechanisms. These different models are not discussed here apart from the concept of a mean repair time and its modifying effect on the duration and postduration of irradiation as used in various forms given by equations 4.10, 4.11, and 4.14, and in Chapter 6 (section entitled *Calculation of Biological Effectiveness in Mammalian Cells*) [Lea, 1955; Harder and Virsik-Peuckert, 1984; Kiefer, 1987].

In general, two-component survival (F_S) is calculated from the combined single hit, single target and single hit, multitarget expressions, equations 4.4 and 4.5, which, because of the influence of the former mechanism, can give the desired initial slope to the survival curve. F_S is of the form

$$F_s = F_{1,1} \cdot F_{1,m} . \qquad (4.15)$$

Although various two-component models have been developed, most of these involve empirically determined parameters, usually a function of LET, in the exponents of the terms in equation 4.15 [Wideroe, 1966; Todd, 1967]. These models are of limited value as the magnitudes of the parameters obtained depend on the radiation type and on the biological endpoint, making it impracticable to achieve a unified system. A common weakness of all models that have multiple floating parameters to be determined by "best fits" to data is that unless the same set of parameters is common to all radiation types, the model remains unproved. The case for validity of the model is greatly strengthened if the magnitude of the determined parameters can be deduced independently from known, more fundamental, biophysicochemical quantities.

Katz's Model for Cellular Inactivation by Heavy Ions

The most successful quantitative formulation of a generalized two-component model of cellular inactivation by accelerated heavy ions has been achieved by Katz [Katz, 1987; Katz et al., 1972]. Although independently developed, there are clear parallels conceptually with Lea's approach in segregating the main ion track and the delta-ray penumbra and then combining the effects of each. As with Lea's model, a major feature of the Katz model is that it is explicitly related to detailed aspects of the physical track structure. The biological response to delta rays is assessed by comparison with the radiosensitivity of gamma rays (and is called "gamma kill"). It is described by a single hit, multitarget function of dose. The response to the ion track core (called "ion kill") is characterized by cross sections. In the model the delta-ray damage is intrinsically independent of photon energy, justified on the argument that it is the associated equilibrium electron spectrum that causes the damage and that this is approximately independent of photon energy. The justification is a poor approximation as there is, in fact, a significant variation in effectiveness over the range of photon energies between carbon characteristic X-rays (250 eV) and ^{60}Co gamma rays (1.17 and 1.33 MeV), as shown in Chapter 6. The net effect of this approach is to reduce to a minimum the contribution made to the bioeffectiveness by the delta-ray electrons surrounding heavy particle tracks. An advantage of the method is that it implicitly allows for the effects of grazing collisions by the delta rays.

Katz attributes the radiation action to the radial distribution of energy deposition by the delta rays generated by passage of the ion. The radiosensitivity of the biological targets to delta rays is determined by extrapolating the experimental radiosensitivities of gamma rays reported in the literature for the same biological endpoint. Formal rules are established to separate the ion core, which comprises a dense radial distribution of low-energy electrons in the delta-ray penumbra. Thus, at radial distances from the ion path a track core may, or may not, be deemed to exist according to the degree by which the extrapolated energy deposition for gamma rays, $\varepsilon_{\gamma,0}$, does, or does not, exceed the limiting reference value, ε_L, of the energy deposition in individual targets assumed to be water equivalent. ε_L has a value of about 1.36 eV/mm. If a track core is found to exist, then there will be a pure exponential survival component (equation 4.4).

Beyond the track core, cumulative action of the delta rays can occur and is described by single hit, multitarget kinetics (equation 4.5). Having segregated the track structure in this way, it remains to calculate the effect cross section for the track core region and the fraction of the total dose responsible for cumulative damage in the track penumbra.

For cells, the effect probability, $P_i(t)$, is calculated as a function of the radial distance t for single hit, multitarget action; i.e.,

$$P_i(t) = (1 - e^{-h_i})^m \qquad (4.16)$$

where the number of hits, $h_i = \varepsilon_i(t)/\varepsilon_{\gamma,0} \cdot \varepsilon_i(t)$ is the mean energy deposition by delta rays in an individual target at distance t. By integrating functions of t to the maximum extremity, T, of the delta rays, the partial action cross section is obtained:

$$\sigma = \int_0^T 2\pi t \cdot P_i(t) \cdot dt \qquad (4.17)$$

and the number of hits, in noncumulative damage, $h_{nc} = \sigma \cdot \Phi$. Substitution into equation 4.4 gives the surviving fraction:

$$F_i = e^{-h_{nc}} \quad . \qquad (4.18)$$

The fraction of cells surviving the cumulative mode of damage is (using equation 4.5):

$$F_\gamma = 1 - (1 - e^{-h_\gamma})^m \quad . \qquad (4.19)$$

To determine h_γ, the limiting value P_L of $P_i(t)$ is used where:

$$P_L = \left(1 - e^{-h_L}\right)^m, \quad h_L = \frac{\varepsilon_L}{\varepsilon_{\gamma,0}} \qquad (4.20)$$

is the fraction of reduced dose, ε_D, deposited in the noncumulative damage mode and therefore,

$$h_\gamma = (1 - P_L) \cdot D/D_{\gamma,0} = (1 - P_L) \cdot \varepsilon_D / \varepsilon_{\gamma,0} \qquad (4.21)$$

with $\varepsilon_D = D \cdot \beta_i^2 \cdot a_0^2 / z^{*2}$ and $\varepsilon_{\gamma,0} = D_{\gamma,0} \cdot \beta^2 \cdot a_0^2 / z^{*2}$. Here the dose reduction term, $b_i^2 \cdot a_0^2 / z^{*2}$, is proportional to the geometrical area of the target per delta ray per unit track.

The net surviving fraction of cells, F_D, is the product of the two components, $F_D = F_i \cdot F_\gamma$.

$$F_D = e^{-h_{nc}} \cdot [1 - (1 - e^{-h_\gamma})^m] \qquad (4.22)$$

where $h_{nc} = \sigma \cdot \Phi$ and $h_\gamma = (1 - (1 - e^{-hL})^m) \cdot \varepsilon_D / \varepsilon_{\gamma,\,0}$.

The square of the ratio of the effective charge on the ion to the ion velocity, z^{*2}/β^2, comes from the Bethe theory of collision stopping power; i.e., the LET, L, for fast ions, which can be written in its approximate form as

$$L = K \cdot \frac{z^{*2}}{\beta^2} \left\{ \ln\left(\frac{2mc^2\beta^2}{(1 - \beta^2)I} \right) - \beta^2 \right\} \qquad . \qquad (4.23)$$

K is a known constant for a specified material [ICRU 37, 1984]. The quality parameter z^{*2}/β_i^2 is of key importance. It is proportional to the yield of ionizations per unit track and consequently the reciprocal, β_i^2/z^{*2}, represents a good approximation to the mean free path for production of delta rays per unit track length of fast ions [Watt et al., 1994]. It follows that for different fast particles at the same velocity, there will be an increasing yield of delta rays as the atomic number of the ion increases, leading to a denser track core. On the other hand, as the maximum energy (and hence range) of the delta ray is given by the term $2\,mc^2\beta_i^2$, β_i^2, the square of the velocity of the ion in atomic units, is a measure of the spatial distribution of the delta-ray penumbra. It is a natural consequence of the model that the effect cross section is able to exceed significantly the geometrical cross section of a single target, which for mammalian cells does not always accord with observation. Saturation effects are fully taken into account via the use of Poisson probabilities. However, the magnitude of the saturation cross section determined from the model is abnormal, possibly because of the approximate treatment of the radiosensitivity to delta rays. Indeed, although the model is very much a delta-ray theory of track structure, minimal radiosensitivity is actually allocated to the delta rays by using the value for ^{60}Co gamma rays. A more general appraisal of the role of delta rays in radiation damage is given in Chapter 5.

The survival dose-response curves are always predicted to have a constant slope (equation 4.6) at high doses, but exceptions to this are observed in practice and sometimes the slope is better simulated by a continuously curving linear-quadratic form. In cellular media the value of the sub-lesion multiplicity, m, varies. It is always greater than or equal to 2, indicating that at least two and often more sub-lesions are required to produce a lesion. However, there are conflicting arguments based on several different independent approaches that the target multiplicity should, in fact, be in the range of 10 to 15 (see Chapter 5). The model explicitly includes the size of the radiosensitive target through a_0, the general magnitude of which is in the micron range and therefore it does not support the widely held opinion that dsbs in the DNA are the key fundamental lesions [Chadwick and Leenhouts, 1981; Chadwick et al., 1992]. Being a single-track model having an important damage component associated with the delta-ray penumbra, there is no scope for coping with dose-rate effects due to multiple tracks. Nevertheless, overall the model achieves a degree of success in describing and predicting the likelihood of induction of various biological endpoints for different radiation types. Part of the success is doubtlessly due to the flexibility obtained by the existence of four floating quality parameters that have to be obtained by "fitting." Katz reasonably justifies the claim that once these have been determined for a specified biological endpoint, then the effectiveness of a wide range of radiation fields can be predicted [Katz, 1994]. The fact that the parameters cannot be determined via independent avenues but must be extracted from experiment is a weakness common to most models. The strength of the track-structure model lies in its global applicability, having been applied to physical, chemical, and biological one-hit detectors and over 40 sets of data for cell survival, mutation, and transformation. It has been used to predict the response of beams of accelerated ions, neutrons, and mixed radiation fields [Katz, 1994]. This is a creditable performance, especially as there is no provision for biological structure other than the target size, a_0, and no allowance is made for repair of damage. In practice, the achievements go beyond that of most other models, yet there are still doubts about its validity [Watt et al., 1994; Goodhead, 1989] and about the feasibility of specifying the response function of practical instrumentation for use in a working system of radiological protection.

Dual Action Models

Dual action models are typically based on the premise that damage may be caused by linear (intratrack) action and quadratic (intertrack) action. In linear action, a single track produces two sub-lesions in the vicinity of each other so that they may interact to constitute the lesion. In quadratic action, each of the two sub-lesions is produced by separate tracks prior to formation of the lesion. The intratrack action is taken to be proportional to dose and the intertrack action is proportional to dose squared. Thus, the yield of lesions is written as

$$Y = \alpha \cdot D + \beta \cdot D^2 \qquad (4.24)$$

and the survival curve is given by:

$$ln(F) = -\left[\alpha \cdot D + \beta \cdot D^2\right]$$
$$F = \exp\left[-\left(\alpha \cdot D + \beta \cdot D^2\right)\right]$$

(4.25)

Equation 4.25 immediately satisfies the requirements of Poisson statistics for the discrete and random energy deposition events and leads to survival curves that have a continuously curving slope at high doses in accord with some of the experimental findings, but in conflict with the constant slope at high doses obtained with the two-component models. Both forms of survival curve have been observed. Another problem is the difficulty in testing the validity of equation 4.24 as a model. Often this type of equation is used for empirical fits to data that may have nothing whatever to do with radiobiology; consequently, it is necessary to demonstrate the validity by conducting a detailed analysis of the underlying composition of the coefficients α and β. The position is exacerbated in the knowledge that, by making simple approximations, nearly all the proposed models can be reduced to the dual action format of equation 4.24. α always represents the slope of the dose-response curve at low doses. Differences between the dual action models arise mainly in the choice of critical "target" and in the underlying assumptions made in the calculation of the probabilities for the coefficients α and β.

Chromosome Aberrations and the Theory of RBE

Neary [1965] modeled radiation action on the hypothesis that primary damage to a macromolecule, understood to be the DNA, in the cell is normally the direct effect of a single energy-loss event in the macromolecule. The concept is applied to the production of chromosome aberrations and, because of similarities in the induction phenomena, is thought to be applicable also to cell killing. The aberrations are presumed to be caused by the interaction in a site of two separate sections of chromosome, each having a primary lesion. The primary lesion could be caused by a single primary ionization produced by one charged-particle track. Two lesions could be caused either by intratrack action for high-LET radiation or by intertrack action for low-LET radiation. From knowledge of the physical properties of the tracks and the geometrical properties of chromosome structure in the cell nucleus, the probabilities of interaction (and hence the yield of chromosome aberrations) can be deduced. The ratio of one-track to two-track aberrations is found to be independent of the diameter of the chromosome and is determined solely by the LET. The model gives a satisfactory theoretical basis for the RBE dependence on LET for chromosome aberrations while emphasizing that, although the basic aberration process is the same, the dependence of RBE on LET is quite different for the linear and quadratic components of dose. Also, there are fundamental arguments concerning the validity of proposed RBE/LET functions without special reference to particle type because RBE versus LET curves are necessarily specific to particle type, as will be discussed later in Chapter 5.

Molecular Theory of Radiation Biology

This is a dual action model that takes the biomolecular structures better into account. Implicit in the model is that the cell nucleus is the target; the 2-nm dimension of the DNA double helix defines the "sensitive site" within which the interactions must take place. Chadwick and Leenhouts [1973, 1981] select the key lesion as a dsb in the DNA double helix which if not repaired can lead to cell death and other biological endpoints. Unlike the other models discussed here in which the direct ionization plays an important role, damage to the DNA is attributed predominantly to the indirect action of diffusing chemical species released by the charged-particle tracks in the vicinity of the DNA. The dsb may be produced by linear-quadratic dose kinetics; i.e., either by interaction of a single charged-particle track or by interaction of two separate tracks, each of which produces a single-strand break (ssb) in proximity to form a dsb. Calculation of the probabilities of DNA rupture take into account a multitude of modifying factors considered to be important, such as track interaction in the vicinity of the nucleotides that are damaged predominantly by diffusing species, the geometrical structure of the DNA, the time available for repair of dsbs, the absorbed dose and dose rate delivered, the metabolic activity of the cell, and the stage in its cycle.

The average number of DNA dsbs is, in general, given by:

$$N = \alpha \cdot D + \beta \cdot D^2 . \tag{4.26}$$

The number of dsbs produced per cell per unit dose by a single track is:

$$N_{1,1} = \alpha D = 2n\mu k \cdot \Omega k \cdot D . \tag{4.27}$$

n is the number of nucleotide base pairs in the DNA per cell. $2n\mu k$ is the probability per cell per unit dose that a break occurs in one of the strands of the DNA molecule. Ωk represents the probability per first strand break that the second strand is also broken in the passage of the same ionizing particle. The parameter Ω is influenced by the structure of the DNA molecule, notably the distance between the two sugar-phosphate bonds, and by the spatial distribution of energy deposition along the track of the ionizing particle. Ω is therefore dependent on radiation type.

Similar reasoning to evaluate probabilities for two-track action gives the number of dsbs per cell per unit dose as

$$N_{1,2} = \beta \cdot D^2 = 2n\mu k(1 - \Omega k) \cdot n_1 \mu_1 k_1 \frac{D^2}{2} \quad . \tag{4.28}$$

Combining the linear and quadratic yields of dsbs gives the total as:

$$N = 2n\mu k \Omega k D + n\mu k(1 - \Omega k)n_1 \mu_1 k_1 D^2 \quad . \tag{4.29}$$

The α coefficient is found to be proportional to the ratio of the mean energy deposited per "site" to the mean energy per event expended in ionization and is therefore almost independent of the type of particle track. The β coefficient is proportional to the site volume per mean energy per event expended in ionization—and is therefore dependent on the type of charged particle present. Chadwick and Leenhouts conclude that cell survival and the induction of DNA dsbs should be a linear-quadratic relationship. Also, there should be a unique linear relationship between the logarithm of survival and the number of DNA dsbs, independently of the way in which the breaks are produced. Allowance is made for repair. Repair of ssbs is considered to be an accurate process, whereas double-strand repair may be unsound and could lead to genetic changes.

The probability for cell survival is:

$$S = \exp(-p_o f_p N) = \exp[-p(\alpha D + \beta D^2] \quad . \tag{4.30}$$

p_0, a constant of the cell type, is independent of dose. $p = p_0 f_p$ is a function of the repair of DNA dsbs. Equation 4.30 gives the cell survival determined by unrepaired dsbs, but the determination of the probability parameters is not straightforward.

Microdosimetry: Distance Model of Dual Radiation Action

Microdosimetry deals specifically with the stochastical distributions of energy depositions by charged-particle tracks in microscopic volumes of cellular and sub-cellular dimensions with the objective of interpreting biophysical damage mechanisms. The model has evolved from the original site model [Kellerer and Rossi, 1974] to the generalized "distance" model [Kellerer and Rossi, 1978]. In the generalized distance model, sub-lesions produced at loci are assumed to occur with a probability that is proportional to the energy, ε_i, transferred at point i. As two sub-lesions are required to produce a lesion, the availability of neighboring loci for interaction, $s(x)$, is given by the product of the sensitive volume and the probability density of any two points at distance x randomly chosen in the matrix. To deduce the number of loci that suffer sub-lesions, it is necessary to know the frequency distribution of energy transfers, $t_D(x)$, which are spaced closely enough to affect these loci. $t_D(x)$ represents the initial distribution pattern of energy transfers. To conform with the dual action criterium, $t_D(x)$ consists of two components: an intratrack contribution producing a distribution $t_D(x)$ of correlated energy transfers that depends on radiation quality but not on absorbed dose, and an intertrack contribution due to a distribution of uncorrelated energy transfers from separate tracks that is independent of radiation quality but proportional to absorbed dose. For spherical shells of radius x around the arbitrarily selected point of energy transfer, the frequency distribution is

$$t_D(x) = t(x) + 4\rho x^2 \cdot r \cdot D \tag{4.31}$$

where ρ is the density of the irradiated medium. Kellerer deduces that the mean probability p for combination of sub-lesions is given by

$$\rho = c \int_0^\infty \frac{g(x) \cdot t_D(x) \cdot s(x) dx}{4\pi x^2} \tag{4.32}$$

and that the number m of sub-lesions is given by

$$m = c \cdot \rho \cdot V \cdot D \quad . \tag{4.33}$$

V is the average volume of the sensitive matrix. $g(x)$ is the distance probability that pairs of sub-lesions will interact. As the mean yield of lesions $\varepsilon(d) = 0.5 \cdot m \cdot p$, substituting values for m and p and expanding $t_D(x)$ gives

$$\varepsilon(D) = c^2 \cdot \rho \cdot V \cdot D \left[\int_0^\infty \frac{g(x) \cdot t(x) \cdot s(x)dx}{4\pi x^2} + \rho \cdot D \int_0^\infty g(x) \cdot s(x)dx \right] \qquad (4.34)$$

$$= \kappa \cdot \left(\zeta \cdot D + D^2 \right)$$

with

$$\kappa = \frac{c^2 \cdot \rho \cdot V \cdot D}{2} \int_0^\infty g(x) \cdot s(x)dx$$

$$\zeta = \int_0^\infty \frac{g(x) \cdot s(x) \cdot t(x)dx}{4\pi \cdot \rho \cdot x^2} \bigg/ \int_0^\infty g(x) \cdot s(x)dx \ . \qquad (4.35)$$

ζ, a property of the medium, is the dose average specific energy density and is proportional to the number of neighboring sub-lesions around a randomly selected sub-lesion. At a dose ζ the linear component equals the quadratic component. As is to be expected, intratrack action dominates at lower doses and intertrack action is more important at larger doses. This distance model reduces to the earlier site model if the loci of sub-lesions are randomly dispersed over a spherical site of diameter, d, and if the interaction probability, $g(x)$, is made constant.

Limitations of the Dual Action Models

An immediate difficulty with the distance model is that in seeking generality, it contains no detailed specification of the site functions $s(x)$ and $g(x)$, although it is possible to make intuitive suggestions for these. For example, by appropriate choice of $g(x)$ to enhance the short-range interactions of sub-lesions while leaving the long-range interactions unaffected enables a major criticism of the earlier site model to be accommodated, based on the apparently anomalously high radiation effectiveness per unit dose by low-energy characteristic X-rays [Cox et al., 1977].

Sub-lesions are assumed to be produced with a probability proportional to the energy transfer, ε_i, at a locus. There is no evidence that this is in fact so and, indeed, in Chapter 5 it will be argued that sub-lesions are produced by the interaction independently of the energy transfer.

Although the model is deliberately expressed in very general terms to cope with any likely mechanisms, there are situations for which modification would be required. For example if dsbs in the DNA are key lesions, then it will be necessary to alter the functions $s(x)$ for the availability of loci and $g(x)$ for the distance probability to distinguish paired sub-lesions in a DNA segment from other paired lesions which may have the same spatial distribution but which would not produce a dsb.

Additional problems arise with the quality dependence of κ and ζ. The quadratic component, κ, (or β in equation 4.24) differs from most of the other linear-quadratic models in that it is seen to be independent of radiation quality and z (or the ratio α/β from equation 4.24) is quality dependent. In fact, experimental studies with monoenergetic fast ions [Blakely et al., 1979] show a clear quality dependence of both the α and β coefficients when quality is expressed as a function of LET or the dose-weighted mean of the specific energy, z_D. Furthermore, as LET increases above values of ~180 keV/µm, β reduces rapidly to zero and the dose-response curve becomes linear throughout the dose range. As the intertrack contribution has its origin in the $t_D(x)$ distribution of energy transfers, a negligible β component requires that $\kappa \ll \zeta$ (equation 4.35), the likelihood of which can be determined for a particular biological endpoint by examining the behavior of these terms with respect to each other in equation 4.20. Sedlak [1988] has extended a microdosimetric treatment to take account of the different radiosensitivity of cells.

Resonant Action In Radiation Damage?

Among the difficulties experienced by the generalized dual action model are the problems of the size of the spherical radiosensitive site deduced using ultrasoft X-ray irradiations [Goodhead et al., 1977] and the proof that κ in equation 4.35 was influenced by the radiation quality of accelerated heavy ions and not constant as predicted [Blakely et al., 1979]. In an attempt to circumvent these problems, Yamaguchi and Waker [1982] proposed the novel idea that the physical mechanism of damage may be the result of a resonance between the spatial distribution of the ion clusters produced by the ionizing radiation and the critical elements of the biological system. Encouragement to take this view resulted from the observation of the apparent invariance between the maxima in the RBE versus LET curve at 100 to 200 keV/µm for different chemical and biological conditions. However, the maximum is not invariant for particle type [Cox et al., 1977; Watt et al., 1985]. Three stages are proposed in their resonance model: the resonant physical stage, a chemical stage with response determined by the G value for the Fricke dosimeter (dependent on radical yield), and a biological stage influenced by a repair mechanism. By analogy with the formula for power absorption in a damped Helmholz resonator, replacing the natural and driving frequencies, respectively, by a LET constant of the biological system, L_0, and the radiation LET, L, the following expression is obtained:

$$F(L) = \frac{2kfL^2}{(L_0^2 - L^2)^2 + (2kL)^2} \ . \qquad (4.36)$$

k is a damping coefficient and f is the driving force strength. By assuming $F(L) \cdot \Delta L$ represents the probability for lesions per traversal of the cell nucleus by an ionizing particle having LET between L and $L + \Delta L$, then the number of lesions per unit of absorbed dose, I_p, can be expressed as:

$$I_p = \frac{2krL^2}{(L_0^2 - L^2)^2 + (2kL)^2} \cdot \Delta L \ . \qquad (4.37)$$

where $r = \rho \sigma f / 16$ characterizes the radiation effect of interest. ρ is the density and σ_g, in micrometers squared, is the geometric cross section of the sensitive structure that is taken to be the whole cell nucleus. f is a dimensionless constant and k has the dimensions of LET. The number of lesions, $\varepsilon_p(D)$, at absorbed dose, D, is given by

$$\varepsilon_p = I_p \cdot D \ . \qquad (4.38)$$

ε_c, the number of lesions due to chemical action, is taken to be proportional to the product of the G value and dose, $G \cdot D$, and the repair of the repairable proportion of lesions is assumed to occur at a rate e^{-aT}, where T is the elapsed time since the end of the acute irradiation. The model can be expressed in terms of microdosimetry distributions and quantities. There are four fitting parameters: L_0, r, k, and c, the chemical proportionality constant. A good fit is obtained to results for human fibroblasts [Cox et al., 1977] irradiated by 250 kVp X-rays, ^{60}Co gamma rays, helium, boron, and nitrogen ions. L_0 and k are found to be 208 keV/µm. There is no D^2 term as the quadratic action is attributed to repair. Dsbs are favored as the lesion, but the site size is estimated to be about 12.5 nm. There is no biological evidence that there is a "resonant" signal (i.e., in the sense of supralinear amplification of the normal signal), but there is indication of a maximum signal produced under conditions where the ionizations along a track "match" exactly in a correlated way with the strand spacing in the

DNA (refer to Chapter 5). Testing with more biological systems could be informative.

A "Delta Function" Track Model of Dual Radiation Action by Fast Heavy Ions

Action of Delta Electrons in Fast Heavy Ion Irradiations

As fast heavy ions dissipate the major part of their energy loss in material through their associated delta-ray electrons, it is not surprising that models for their bioeffectiveness should pay special attention to their role. Useful information on the relationship between properties of the electron track structure and the induced lesion can be obtained from the pioneering electron probe studies of mammalian cells carried out by Cole and his colleagues [Zermeno and Cole, 1969; Datta et al., 1976] who measured effect cross sections as a function of electron energy for viruses and cells. Re-examination of this work for induction of reproductive cell death in Chinese hamster cells, using current knowledge of electron track structure, leads to conclusions that differ from those of the original authors and suggests a mechanism of inactivation that is not incorporated in any of the radiation damage models [Watt et al., 1984]. The trend of the inactivation cross section with electron penetration depth is considered in relation to the structural dimensions of the cells by comparing plots of the damage against the distribu-tion of electron penetration ranges, the fractional transmission, and the energy deposition per unit track length corrected for electron straggling. None of these quantities is in accord with the observed trend in the cross sections. However, if comparison is made with the electron slowing down fluence and concentration of electrons with depth, a more reasonably convincing correspondence is found.

Three important conclusions can be deduced from the known physical properties of the electron transport: (1) the lesion is induced by the cascade of electrons produced near the end of the primary electron tracks as they degrade in energy; (2) there is confirmation that the extranuclear region of the cell is radiation insensitive and that the thickness of the insensitive region is equal to the penetration depth of a low-energy electron (10 keV ~ 2.5 µm; 20 keV ~ 8.5 µm) for the specified biological endpoint (cell death); and (3) as the low-energy electrons cascade, and the maximum effect cross section does not develop until the primary electron track has penetrated the cell nuclear membrane, the radiosensitive targets must lie within the cell nucleus. The nuclear membrane appears to play, at most, a minor role.

"Delta Function" Model of Dual Radiation Action

To take into account the conclusions on damage mechanisms for electrons discussed in the last section, Cannell [Watt et al., 1984] has proposed an alternative dual action model. His assumption is that the radiation produces a "soup" of lesions and sub-lesions that are randomly distributed in the site. A lesion may be produced either directly by a single-track event or by the interaction of sub-lesions produced by two or more separate tracks. As the interaction of two sub-lesions is much more probable than higher-order interactions, only the linear and quadratic terms are considered. The linear component due to intratrack action is proportional to dose and associated with the α coefficient. The intertrack action, associated with the quadratic β component, is proportional to D^2. It comprises lesions produced by interaction of time coincident sub-lesions. As some of the sub-lesions will arise from single-track effects, allowance must be made in the overall accountancy. Thus, in general, the linear-quadratic coefficients can be written as:

$$\alpha = a(1 + c \cdot f(L)) \quad \text{and} \quad \beta = [b(1 - f(L))]^2 \qquad (4.39)$$

where a is the dose efficiency conversion factor for production of single-track lesions; b is the efficiency with which the fraction of the dose not expended in intratrack action is converted to lesions by intertrack action; and c is related to the efficiency with which intratrack sub-lesions interact to produce lesions. $f(L)$ is the quality dependent fraction of dose expended in intratrack sub-lesions. To determine the form of $f(L)$, it is assumed that there is an average distance x between sub-lesions and that if the physical mean free path for track interaction λ between energy transfers to the medium is close to x, then production of a sub-lesion is more probable. If λ is greater than or less than x, the probability decreases. $f(L)$ is satisfied by the delta function:

$$f(L) = \frac{1}{\sqrt{\pi \cdot \varepsilon}} \cdot \exp\left[\frac{-(1 - L)^2}{\varepsilon^2}\right] \qquad (4.40)$$

where L is the LET normalized to 180 keV/μm, ε is the width of the distribution, and $1/\varepsilon\sqrt{\pi}$ is the maximum amplitude. Support for this choice of the $f(L)$ function is found in the trend of the linear-quadratic coefficients α and β with LET and in the trend of RBE with LET [ICRU 16, 1970].

Maximum values of α, β, and RBE for fast medium ions seem to occur at LETs near 180 keV/μm. The value of x works out at approximately 30 nm. The magnitudes of both the LET and the site size of 30 nm immediately raises doubts about the validity of the approach, in particular the site size, because it is not readily associated with important biological structure. In the initial development of this model, it was assumed that radiation quality could be adequately specified by the LET, although some function of z^{*2}/β^2 and β^2 may have been better as this would allow explicitly for the yield and spatial distribution of the delta rays. (Refer to Katz' equation 4.23 earlier in this chapter.) Consistent values for the parameters a, b, and c used in equation 4.39 are obtained when the model is fitted to the data of Skarsgard et al. [1967] for irradiation of Chinese hamster cells with fast ions and X-rays. Also, the survival curves calculated using the coefficients are found to give very good fits to the experimental data and tend to pure exponential survival at high LET.

The concept of a "distance" matching introduced a distinct change in interpretation of the maximum of the RBE versus LET curve that can be explained by a "match" between the mean free path between interactions of the radiation and the mean interaction distance between sub-lesions. The conventional explanation is that it is due to competition between increased efficiency and saturation effects due to energy wastage. However, as the position of the maxima RBEs when plotted as a function of LET are particle-type dependent and occur at particle energies at which the LET continues to increase (and hence the mean free path for interaction continues to decrease), it follows that excess ionizations will occur within a relatively small decrement in particle energy, thereby leading to saturation effects (i.e., interaction wastage, $\lambda < 2$ nm) at high LET and masking the "strand matching" effect. Careful examination of effect cross sections in the immediate neighborhood of the onset of optimum effect for individual particle types produces some evidence that this is indeed the case, but the results are not conclusive.

Bond and Varma's "Hit-Size Effectiveness" Model

Bond and Varma [1981 and Varma and Bond, 1987] combine the principles of hit theory with methods of microdosimetry to form their "hit-size effectiveness" theory of radiobiological action in cells or organs. The "hit" constitutes passage of a charged-particle track through a critical volume within which the unspecified radiosensitive targets in the cell are contained. Each time a hit occurs, a stochastic amount of energy will be transferred to the critical volume and may induce the biological endpoint if the amount of energy lies above a certain threshold energy. The severity of the damage depends on the event size. Distinction is made between high dose exposures and low dose exposures. For high dose exposures (≥ 1 Gy) there will be multiple hits of the critical volume, in which case absorbed dose is an appropriate quantity to predict the incidence of the biological effect in an exposed cell population. However, the main interest lies in the low-level radiation exposures (<1 Gy). For these, the stochastics become significant. Only a small fraction of the cells may be hit. The chance that a cell will be hit is best determined from the fluence of charged particles rather than the dose. Thus, the average number of hits per cell is given by $I_H = \sigma_{cv} \cdot \Phi$, where σ_{cv} is the projected cross-sectional area of the critical volume. Because this is an average quantity, the number of hits is proportional to fluence, which, for low fluence, makes it impossible to determine the risk of injury to a single cell. Consequently, stochastical analyses are required. As it is not possible to find the stochastic distribution of hit sizes in actual living cells, the simulated microdosimetric distributions are measured with a spherical proportional counter of site size equivalent to the critical volume. Each hit in the distribution is then weighted by a pre-determined response function, known as the "hit-size effectiveness response," to predict the fraction of cells that suffer the specified biological effect.

The Hit-Size Weighting Function

Information on the charged-particle fluence and the incidence of hit cells can be determined from the microdosimetry distributions and knowledge of the cross section for the critical volume. Radiation quality is determined solely by the energy deposition. However, the result is quality dependent and so it is desirable to obtain a means of applying the method to any radiation quality. This is achieved by first determining the differential distribution of the incidence of hit cells receiving energy transfer, ε, as a function of ε, for low- and high-LET radiation. Next, the fraction of hit cells that suffer the biological endpoint is obtained as a function of energy transferred; i.e., hit-size to cover the range of low- to high-LET values. Combining the two spectra yields the incidence of cells responding as a function of hit size. This latter is the "hit-size weighting function." No provision is made for sub-lethal damage as it is considered to be completely repairable.

Critical Comment on Validity

An important feature of this model is the presumption that the probability of a biological lesion occurring is directly related to the amount of energy transferred (via ε_i or z_D). However, comparison of the energy deposited by heavy charged particles for equal degrees of biological damage shows that this does not happen. The range of LET used, limited to <150 keV/μm, is insufficient to provide a sensitive test of the model. Katz [1994] has criticized the concept on the basis that it does not predict effect cross sections or correlate with observed cell survival data. Simmons [1992] critically discusses application of the model in a fluence-based system of risk control. Waligórski [Waligórski and Olko, 1991] finds that neutron data cannot be used with either the Katz model or the hit-size effectiveness model to predict risk and that therefore casts doubt on the value of these models for radiation protection.

Lethal-Potentially Lethal (LPL) Model

Curtis [1986] hypothesizes that cell death can be caused by either of two main types of radiation-induced lesion, one irreparable and the other repairable. The nature of the lesions is not specifically identified, but it is implicitly assumed that these are types of dsbs in the DNA of different severity. Irreparable lesions have a linear dependence on absorbed dose, D, with a proportionality constant defined by η_L. Repairable lesions are more complex. Two classes of repairable lesion are suggested, categorized according to whether their rates of repair are slow or fast and occurring at rates denoted by ε_L and $\varepsilon_L{}'$, respectively. Breaks in the DNA with long repair times are attributed to those occurring in the linker DNA between nucleosomes, whereas the more rapidly repairing breaks occur in the DNA bound to the nucleosomes. The slowly repairing breaks are thought to be more serious as they can undergo "binary misrepair" by rejoining erroneously with another part of the DNA leading to a fatal lesion produced at a rate ε_{2PL}. As with the irreparable lesions, both classes of repairable lesion are assumed to be formed directly in proportion to the absorbed dose but with different proportionality constants, η_{PL} and $\eta_{PL}{}'$, for the slow and fast components.

Mathematically the model can be embodied in two differential equations to determine the rate of change of the mean numbers per cell of the lethal, n_L, and potentially lethal, n_{PL}, lesions:

$$\frac{dn_L(t)}{dt} = \varepsilon_{2PL} n_{PL}^2(t)$$

$$\frac{dn_{PL}(t)}{dt} = -\varepsilon_{PL} n_{PL}(t) - \varepsilon_{2PL} n_{PL}^2(t) \quad .$$

(4.41)

By applying the initial conditions: $n_L(0) = \eta_L D$ and $n_{PL}(0) = \eta_{PL} D$, the solution to the equation gives the yield of lesions. Then using Poisson statistics, the survival fraction of cells that have a mean repair time, t_r, is:

$$S = \exp\left\{-\left(\eta_L D + \eta_{PL} D\right) + \varepsilon \ln\left\{1 + \frac{\eta_{PL} D}{\varepsilon}\left[1 - \exp\left(-\varepsilon_{PL} t_r\right)\right]\right\}\right\} \tag{4.42}$$

with $\quad \varepsilon \equiv \dfrac{\varepsilon_{pl}}{\varepsilon_{2PL}}$.

Approximations to the LPL model tend to the linear-quadratic form for low doses and low-LET irradiations and to an exponential slope at high doses. Curtis [1989] has demonstrated that the Katz expression also becomes linear-quadratic at low doses and low LET, in which case there are simple relationships between the parameters of the Katz model, the LPL, and linear-quadratic models [Curtis, 1987] thereby formally demonstrating the extremely difficult problem of appraising the validity of different models and the concepts involved. Interestingly, improved agreement with the biological data was obtained when z^2/β^2 was substituted for LET as the quality parameter [Curtis, 1989].

Track Structure Model of Damage to DNA

Chatterjee and Holley [1991] have used a highly sophisticated approach to modeling the production of strand breaks in the DNA by combining the delta-ray structure of the charged-particle track with the structural chemistry of the DNA molecule to investigate damage mechanisms. Monte Carlo methods are applied [Holley et al., 1990].

Assumptions are that the intracellular DNA is the critical target and that damage to the DNA can possibly lead to cell death, transformation, and mutations. Special attention is given to the chemical effects of diffusing radicals (indirect effects) produced by radiation interaction in the intracellular water and to the direct effects caused by the physical ionization produced along the charged-particle tracks. The hydroxyl radical (OH) is generally recognized as being the most significant. It requires an expenditure of approximately 17 eV of energy to produce a hydroxyl radical in water. It has a diffusion length of about 3 nm in cellular material with a reaction distance of approximately 0.1 nm with sugar and 0.4 nm with the bases in DNA. If OH radicals reach the DNA, they can produce strand breaks only by extracting hydrogen from the sugar moiety of the strand. Strand breaks are not produced in interactions with the bases, only point mutations. The calculation of the production of DNA strand breaks involves relating details of the track structure to the structure of the DNA molecule. It proceeds first by dividing the heavy particle track into a "core" and "penumbra" (delta ray) region [Chatterjee, 1987]. A three-dimensional record is made of the structure of the DNA, using X-ray diffraction data, and containing the coordinates of each sugar molecule and each base molecule. Monte Carlo techniques are applied to follow the initially Gaussian distribution of diffusing radicals to determine the yield of strand breaks. From this, separate values of the D_{37} can be calculated for ssb production by indirect effects associated with the core and penumbra components of the track. Results are expressed as cross sections for production of ssbs, σ_{SSB}, given by:

$$\sigma_{SSB}\left(cm^2\right) = 1.6 \cdot 10^{-9} \cdot L \cdot \left[\frac{f_{core}}{D_{37}^{core}} + \frac{\left(1 - f_{core}\right)}{D_{37}^{pen}} \right] \qquad (4.43)$$

with L in keV/μm, D in gray, and f_{core}, the fraction of energy deposited in the core.

Direct effects are determined by calculating the average energy deposition due to excitation and ionization of the DNA sugar-phosphate backbone by fast-charged particles. The consequent deprotonation or direct dissociation are assumed to cause ssbs. Those ssbs that occur within 10 base pairs of each other, and on opposite strands, are presumed to lead to dsbs. Thus for an incident beam of heavy particle fluence, Φ cm^{-2}, if the calculated number of ssbs and dsbs are, respectively, N_{SSB} and N_{DSB} in DNA of molecular weight M daltons, then the cross sections for ssbs and dsbs are, respectively:

$$\sigma_{SSB} = \frac{N_{SSB}}{M \cdot \Phi}$$

$$\sigma_{DSB} = \frac{N_{DSB}}{M \cdot \Phi} \qquad (4.44)$$

which gives the normalized production cross sections for ssbs and dsbs in cm^2 per dalton. The cross sections can easily be expressed as yields per unit dose in gray by multiplying the cross sections in equation 4.44 by $6.25 \cdot 10^8 \cdot \rho/LET$ (keV/μm), density ρ in g/cm^3.

Considering that the theory is based on an artificial biological system (viz. aqueous solution) and the DNA used is of linear form in a small virus, remarkably good agreement is achieved in comparison with experimental data for mammalian cells. Excellent agreement is obtained in the D_{37} values, attributable solely to indirect effects, between experiment and theory for fast electrons and accelerated ions in the aqueous solution containing 10 mM of radical scavenger.

Extension of the calculations to the breaks observed in the chromatin form of DNA can also be plausibly explained on the basis that only breaks in the linker sections of this form of DNA are detected. For ssbs, indirect and direct effects occur with approximately equal probability and decrease with increasing LET. For dsbs, the mean energy expended per break is about 60 eV and a maximum in yield of dsbs is observed at about 220 keV/μm.

No repair is involved in the examples cited but the success achieved gives confidence in the conceptual basis of the damage mechanisms that warrant fuller investigation for other radiation types and cells. The arguments will be further strengthened if it can be demonstrated that the constants in the model are indeed universal constants as claimed.

Later developments of the model include allowance for specific types of damage in the DNA molecule that are thought to be precursors leading to radiation-induced mutagenesis and carcinogenesis. Very encouraging results are obtained for the heavy ions but less so for the fit to the X-ray data. In the latter context, it is interesting to note that the LET for 225 kVp X-rays is taken to be 0.9 keV/μm, which is about a factor of five and a half times too small for electrons in the equilibrium conditions pertaining.

General Comments on Modeling

Consideration of the limited degree of success achieved by the modeling approaches described previously, and the diverse nature of the basic concepts and quality parameters involved, promotes speculation on the validity of the proposed physical mechanisms of radiation action. Key parameters used as descriptors of the radiation quality are: the primary ionization density (Lea), which is proportional to the restricted LET (Harder); the unrestricted LET; track and dose average restricted LETs; z^{*2}/β^2 (Katz), which is a dimensionless quantity proportional to the linear primary ionization and/or to the delta-ray yield per unit path length of heavy particle track; the microdose quantities: track average lineal energy, y_F, which is proportional to LET for fast heavy particles, and the specific energy density, z_F, which is proportional to absorbed dose for heavy particles [ICRU 36, 1983]. A possible reason why simple tests cannot be applied in practice to determine the respective merits of these quantities is that most of the models have arbitrary constants associated with the quality parameters and, as can be seen in Figures 4.2(a-e), there is an approximate proportionality among the various quantities over a large range of particle energies. The combined effect makes it difficult to distinguish significant features when data are modeled using fitting techniques and compared on logarithmic plots.

Another difficulty in testing models results from the very poor "signal to background noise" ratio and the consequently low statistical precision of measurement. The following hypothetical example can illustrate just how poor is the statistical precision.

Assume that a charged-particle track traversing a 7-micron mean chord length through the cell nucleus produces paired bond-breaks in each of 10 sensitive sites (e.g., the intracellular DNA) to cause inactivation. Then the actual energy expended in bioeffectiveness (i.e., the signal) is overestimated to be about 0.2 keV, but the energy deposited in the cell nucleus is, on average, seven times the LET in keV/μm. The difference in energy is the "background noise." Table 4.1 shows that the ratio ranges from 10^{-2} to less than 10^{-4}. Therefore, to assess the respective merit of energy-based quality parameters, it is necessary to compare the net "signals", which at the levels quoted here, make the exercise virtually meaningless unless the background noise can be substantially reduced. A practical observation of the consequences is that the contribution due to delta rays, to the action cross section for inactivation in the saturation region by argon ions, in a fluence-based system, may extend to 300% of the saturation level and is therefore very significant. Yet this large damage contribution is not observable as a change in RBE in the dose-based system (see *The Role of Delta Rays Associated with Fast Heavy Ion Tracks* in Chapter 5 and Figure 5.5).

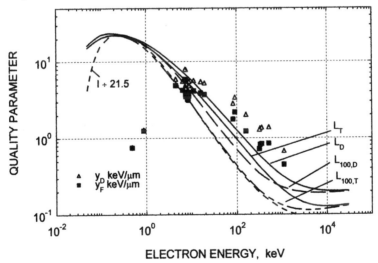

Figure 4.2(a). *Values of the quality parameters, track, and dose average LET; the 100-eV restricted LETs; lineal energies all in keV/μm; and the linear primary ionization (normalized to the track average LET at 0.3 keV/μm) are shown for the equilibrium spectrum generated by initially monoenergetic electrons in water. [The physical data shown in Figure 4.2(a-e) are taken from Watt, 1996.]*

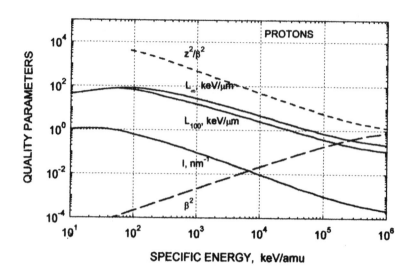

Figure 4.2(b). *Quality parameters for water are shown for track segment experiments using protons.*

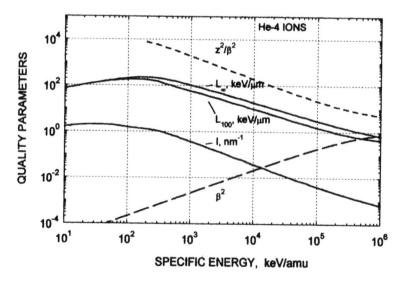

Figure 4.2(c). *Quality parameters for water are shown for track segment experiments with helium ions.*

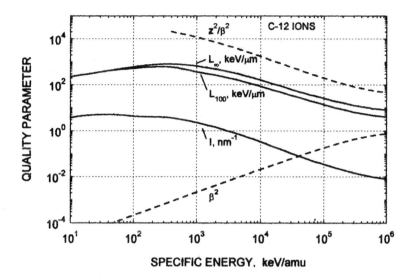

Figure 4.2(d). *Quality parameters for water are shown for track segment experiments using carbon ions.*

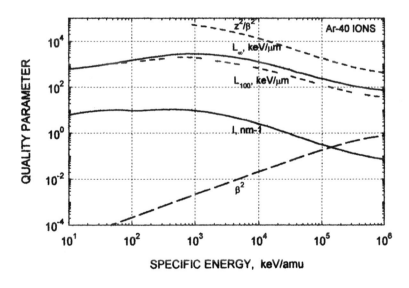

Figure 4.2(e). *Quality parameters for water are shown for track segment experiments using argon ions.*

Table 4.1. *Signal to background energy ratios for optimum inactivation**
of mammalian cells.

Radiation Type and Energy	Average Energy Expended per Nuclear Traversal	Signal to Background Ratio
Electron; 20 keV	20 keV	1.0×10^{-2}
Neutron; 850 keV	362 keV	5.5×10^{-2}
Proton; 370 keV	434 keV	4.6×10^{-4}
Helium-4; 2.92 MeV	903 keV	2.2×10^{-4}
Lithium-7; 12.3 MeV	1.15 MeV	1.74×10^{-4}
Carbon-12; 78 MeV	1.59 MeV	1.26×10^{-4}
Neon-20; 360 MeV	1.93 MeV	1.04×10^{-4}
Argon-40; 2400 MeV	2.38 MeV	8.0×10^{-5}
Iron-56; 7784 MeV	2.75 MeV	7.27×10^{-5}
Kr-84; 2.88×10^4 MeV	3.10 MeV	6.45×10^{-5}

* Optimum inactivation is at the onset of the saturation cross section, at which point energy wastage is a minimum, the RBE is a maximum, and the mean free path for linear primary ionization is ~2 nm.

5.

Radiation as a Probe for the Physical Investigation of Radiosensitive Structure in Biological Targets

The advantages to be gained by utilizing radiation probes in the study of nuclear, atomic, and molecular structure have long been recognized. This is also true in the study of radiation effects in cell biology. Typically, the net biological effectiveness of the radiation field is measured either directly in the form of effect cross sections or as dose-survival curves. The probes are the charged-particle tracks that are present in the immediate vicinity of the radiosensitive sites. For photon or neutron irradiations, the relevant charged-particle probes are, respectively, the electron fluence at equilibrium and the recoil proton fluence. In track segment experiments, the probes may be beams of accelerated ions or electrons. Cole et al. [1974] and Datta et al. [1976], using electrons or alpha particles, have explicitly applied the technique to determine the depth of sensitive structures in the cell. Biophysicists working with accelerated ion beams typically study the trends of cross section with ion energy or with quality parameters, usually LET, in an attempt to deduce energy-dependent mechanisms of radiation effect. Hitherto, none seem to have applied the power of the method specifically as an investigative tool to explore nature, structure, and distribution of radiosensitive sites within the cell nucleus.

The first step in the procedure used here is to determine the cross section for production of the biological effect. This cross section represents the cumulative probability per target (e.g., cell nucleus or specified radiosensitive site) per unit fluence of the relevant charged particles for production of the biological endpoint of interest. It is an absolute measure of the average bioeffectiveness of the ionizing radiation. Consequently, if cross sections for the induction of the biological endpoint can be obtained, then the trends as a function of the spacing of interactions can be examined to identify and reveal characteristic features of the radiosensitive sites and their distribution within the cell.

Sometimes the actual cross sections rather than the survival fractions are reported in the literature as, for example, in experiments with accelerated heavy particle beams. Otherwise, the cross sections have to be extracted from the dose-response curves. This is usually the case for irradiations with indirectly ionizing radiations.

The procedure for determining the effect cross section assumes that the survival fraction can be written as a pure exponential of a function, usually linear or linear-quadratic, for the yield of lesions. Then, if the fractional survival, F, is observed at a point where the charged-particle fluence is ϕ_p, the effect cross section is:

$$\sigma_B(\phi,t) = -\frac{\ln F}{\phi_p} \qquad (5.1)$$

where σ_B may be a function of fluence and time. For a dose-response curve, the cross section is determined by:

$$\sigma_B\left(cm^2\right) = 1.6 \cdot 10^{-9} \cdot \frac{\overline{L_T}\left(keV / \mu m\right)}{D(Gy) \cdot \rho\left(g / cm^3\right)} \cdot \left(-lnF\right) \quad . \qquad (5.2)$$

D is the absorbed dose in gray. $\overline{L_T}$ is the track average LET weighted for the relevant charged particle spectrum generated by the radiation field and ρ is the density of the biological medium. For convenience the cross sections are determined usually from the initial slope of the dose-response curve (i.e., from $-\partial(\ln F)/\partial D$ as D→0 and damage is assumed to be prior to repair); from the D_{37} (at which point survival is $1/e$ and there is, on average, a single, $-\ln(1/e) = 1$, "hit" per target) or from the final constant slope (if it exists). In the illustrative calculations of cross sections made here, the initial slope is used.

However, for quantitative comparison of the radiation effectiveness in a wide range of biological material such as enzymes and some types of bacterio-phage known to have single-hit response and other types of bacteriophage and mammalian cells known to have a more complex response, it is more meaningful to express the effect cross sections as ratios to the projected geometrical cross-sectional areas of identifiable structures; e.g., of the whole enzyme, cell nucleus, etc. This permits the possibility of unifying the data for any target type and for any radiation type. Later it will be argued that the projected area of the DNA content in the cell nucleus combined with the mean number of DNA segments (targets) at risk along a mean chord traversal is the appropriate unifying quantity because this quantity constitutes the saturation cross section for induction of the biological effect. These ratios represent the average cumulative probability, or intrinsic efficiency, for induction of the biological endpoint by an individual track in the relevant charged-particle spectrum. It is possible for the cumulative probability to exceed unity as the observed damage ratios necessarily include the sum effects of the direct (ionizations) and indirect (diffusing water radicals, etc.) radiation actions as well as the contributions from any delta rays that may be associated with the track of an accelerated ion. This has been very clearly demonstrated in experiments on single-hit targets such as enzymes where the effect cross-section ratio may extend to many times the geometric [Dertinger and Jung, 1970]. It also occurs in mammalian cells, but to a much smaller degree because of the two-hit nature of the inactivation mechanism, as can be seen in the saturation region for heavy particle inactivation of mammalian cells shown in Figure 5.1.

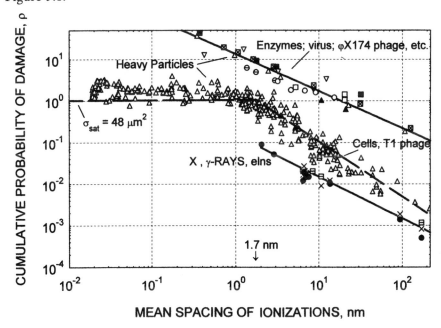

Figure 5.1. *Cumulative probabilities for inactivation are shown for some enzymes, viruses, bacteriophage, and mammalian cells for irradiations with accelerated ions, X-rays and gamma rays, and electrons. The objective is to identify any internal structure that may be present in mammalian cells with a view to exploring methods for unifying damage data for a diverse range of biological targets. For further discussion, see* **Interpretation of Damage Mechanisms** *later in this chapter. [See* **Source of Data for Figures** *in References for Chapter 5.]*

Study of the trends of the observed cross-section ratios with proposed quality parameters has the potential to reveal information on structural details of the radiosensitive sites within the target, the nature of the damage mechanisms involved, and the suitability of the quality parameter(s) for quantifying the radiation action. Quality parameters selected for test (because of their relevance to modeling) are the mean free path for linear primary ionization, restricted LET with 100-eV cutoff, and the LET of tracks in the charged-particle equilibrium spectrum. The parameters z^2/β^2 and the frequency mean lineal energy are covered indirectly. This is because z^2/β^2 applies only to fast particles; but it is directly proportional to the more fundamental quantity, the linear primary ionization that applies over the whole energy range. The microdose quantity, the frequency mean lineal energy, is almost identical to the LET for fast heavy particle traversals of a sphere if delta-ray escape is neglected. See, for example, Appendix A in ICRU Report 36 [1983]. The main results for the cumulative probabilities of inactivation for an extensive range of data taken from the literature are shown in Figures 5.2(a), (b), (c). Graphs in Figures 5.2(b) and (c) are plotted against the reciprocal LETs rather than against the LETs to reverse the curves and facilitate direct comparison with Figure 5.2(a). The scatter of the data points, thought to be due to biological, physical, and experimental variability, makes it difficult to decide on the respective merits of the three quality parameters tested. Nevertheless, the 64 data points in the nonsaturation region to the right of the point of inflection at 1.8 nanometers (nm) in Figure 5.2(a) is better correlated statistically to a curve of the form $y = a(1 - \exp(-x_0/x))^b$ as shown in Table 5.1. x represents the selected quality parameter and x_0 represents the reference mean value. Bearing in mind that the mean free path for linear primary ionization, λ (lambda), is the more fundamental quantity, being the zeroth moment of energy transfer, compared with the LETs that are first or higher moments if dose weighted, the choice of lambda should be favored as the best quality descriptor.

There are other arguments in support of this choice. First, lambda permits more explicit interpretation of mechanisms. The point of inflection at ~2 nm is meaningful because of its connotation with regard to target sizes and lesions in the DNA, independently of radiation type, which permits correlation of the cumulative probabilities for cell inactivation into a single (unified) curve by expressing the probabilities with respect to the saturation cross section given by $\sigma_{g, DNA} \cdot n_0$. Second, consideration of other related factors to be discussed in the following sections gives enhanced confidence that the lambda of charged-particle tracks is the main characteristic that determines their biological effectiveness by direct interaction. Lambda represents the number and spacing

Radiation as a Probe for the Physical Investigation
of Radiosensitive Structure in Biological Targets

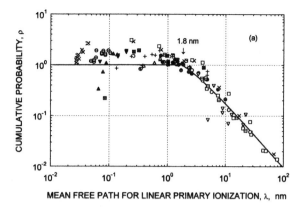

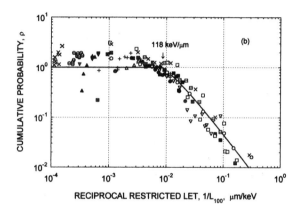

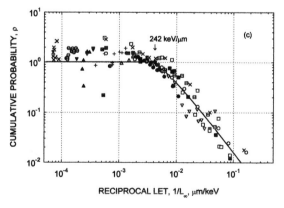

Figure 5.2. *Cumulative probabilities for inactivation of Chinese hamster V-79 cells are shown as functions of the quality parameters, λ, 1/L$_{100}$, and 1/LET for heavy accelerated particles. [See **Source of Data for Figures** in References for Chapter 5.]*

Table 5.1. *Fitting coefficients for the unsaturated inactivation data (64 data points) shown in Figure 5.2 are listed for the quality parameters* $x = \lambda$, $1/L_{100}$, *and* $1/L_{\infty}$.

Quality, x =	λ, nm	$1/L_{100}$, μm/keV	$1/L_{\infty}$, μm/keV
S.D. of fitted curve %	5.13	5.53	5.81
$\rho = a(1-\exp(-x_0/x))^b$			
Arithmetic mean deviation of individual points, %	32.9	35.7	37.3
$a =$	1.8	1.9	1.96
$b =$	1.3	1.5	1.5
$x_0 =$	1.8	8.5×10^{-3}	4.13×10^{-3}

of initial energy transfer interactions. It is independent of the magnitude of the energy transfer and hence LET and absorbed dose are invalid.

Using currently available radiobiological techniques, it is likely that experiments designed specifically to measure the cumulative probabilities could be performed with much better statistical precision and lead to a clear resolution of the matter. Typically, the spread of the data shown in Figure 5.1 and listed in Table 5.1 is attributed to biological variability in cell sizes, to limitations in the assay procedures, and to the contributions associated with the duration of the irradiations. Further improvements could be made in the design of the physical aspects of the experiments. For intercomparison of results and interpretation of mechanisms it is important to know the actual cell dimensions during irradiation as they may be flattened in plating or become more spherical in suspension [Braby, 1995] as the geometry affects the multiplicity of targets at risk along a chord length within the cell. Track segment experiments with heavy particles should be conducted under conditions that ensure charged-particle equilibrium for the associated delta rays. The packing density of cells, which influences the number of cells at risk to the track structure, could be important in assessing the influence of secondary electrons. In plated cells, homogeneous monolayers are desirable. Superimposed cells could lead to erroneous interpretation of damage. Similarly, the density of cells in suspension can alter the degree of damage caused by secondary tracks and by diffusion of chemical species. Comparison of experimental measurements of damage yields obtained for pure, plated DNA molecules with the results obtained for the same endpoint in intracellular DNA irradiated in cells must make allowance for the modifying effect of the different spatial distribution of the DNA target molecules, particularly for fast heavy particle irradiations.

The use of radiation tracks as biological probes proves to be a powerful analytical method that provides a host of information concerning the broad nature of the radiosensitive targets, the type of critical lesion involved, and its mode of production. Also, conclusions can be reached on: (1) the existence of a multiplicity of targets within the cell nucleus; (2) the mean number of targets at risk for a charged-particle track traversing a mean chord through the nucleus; (3) the extent of the contribution to the damage due to delta rays associated with heavy particle tracks; (4) an alternative reason why electrons and photons (X- and gamma rays), including Auger electron emitters and ultrasoft characteristic X-rays, are much less damaging per unit fluence than are the heavy particles; (5) the existence of a common causal mechanism for occurrence of the maxima in RBE observed for various ion types and biological endpoints; and (6) an explanation for the apparent anomalies in the quality dependence of the maxima RBEs observed for protons and neutrons when compared with other ions as a function of LET. The use of radiation tracks as biological probes also provides a basic biophysical reason for the magnitudes of the quality factor and radiation weighting factors. A corollary of the results is that absorbed dose is an inappropriate parameter for quantifying radiation effects.

Extending the analyses to other cellular endpoints such as chromosome aberrations, transformations, and certain types of mutation demonstrates that all the endpoints appear to be linked to the same general type of lesion. Scaling ratios can be determined that enable their probability of occurrence to be obtained from inactivation. The probabilities are also of direct importance in radiotherapy.

Interpretation of Damage Mechanisms

Biophysical Argument that dsbs in DNA Are the Critical Lesion

Reference to Figure 5.1 shows that the cumulative probability for inactivation of enzymes and ΦX-174 phage decreases with increasing mean free path, without any characteristic structured features, as may be expected for targets known to have single-hit kinetics. However, the mammalian cell group has two distinct differences. First, the cumulative probability for inactivation is observed to be an order of magnitude less than that for the single-hit targets, indicating a more complex hit-target mechanism for damage in the mammalian cells and T1 phage. Second, structure is observed in the form of a pronounced change of slope occurring at a mean free path of about 1.7 ± 0.3 nm (loosely taken to be 2 nm), a dimension that one automatically associates with the mean chord distance across a segment of the double-stranded helix of the DNA molecule. Support for this interpretation is found in the positions of the ΦX-174 phage, which is a single-stranded DNA version and lies in the enzyme group, compared with the position of T1 phage containing double-stranded DNA, which is grouped with the mammalian cells. Viruses for Newcastle disease [Wilson and Pollard, 1958; three data points] and influenza A [Jagger and Pollard, 1956; two data points] have single-stranded DNA in a helical nucleocapsid and, as may be expected, lie in the single-hit group containing the enzymes. Other materials which are not shown in Figure 5.1 but that contain various forms of double-stranded DNA conform with the mammalian cell group. These are *Escherichia coli*; *Bacillus subtilis* spores [Baltschukat et al., 1986], and *Saccharomyces cerevisiae* (yeast) cells [Kiefer at al., 1982]. The point of inflection near 2 nm always occurs in targets containing double-stranded DNA. Comparison of the nature of the phages and the mammalian cells tends to exclude the possibility that the dominant critical site is in the cell membrane or that it is a region that occurs jointly between the membrane and DNA [Alper, 1979]. The conclusion, from this biophysical evidence, that the key lesions are dsbs in the intranuclear DNA is in accord with the extensive radiobiological opinion formed years ago [e.g., Neary, 1965; Chadwick and Leenhouts, 1973; Bryant, 1985; Frankenberg, 1994], but the interpretation is conceptually different. Further support for the argument that the dsb in DNA is the fundamentally important lesion and that there can be a multiplicity of DNA segments at risk will be presented later in the context of the projected cross-sectional areas at the level of saturation damage observed for electrons compared with those for heavy particles.

Target Multiplicity in a Single-Cell Nucleus

If multiple radiosensitive targets are dispersed within the cell nucleus, then the effect cross section should begin to decrease as the range of the ion probe becomes less than the mean chord distance through the cell nucleus. Support for this argument can be found in Figure 5.3 for protons that have a maximum effect cross section of about 46 μm^2 at a mean free path of ~3.5 nm and range = 4.7 μm; neutrons (proton recoil spectrum, σ_{max} = 19 μm^2 at λ = 1.9 nm; range ~4.7 μm) and for Fe-56 ions at the saturation cross section, (σ_{sat} = 40 μm^2 at λ = 0.082 nm; range = 5.8 μm). The maximum delta-ray range for Fe-56 ions at its maximum cross section is about 25 keV, which is sufficient to explain the excess cross section above the saturation value. At shorter lambda values, the effect cross section is seen to decrease as the ion range becomes less than the cell nuclear dimension for protons (p), neutrons (n), and the Fe-56 data marked on Figure 5.3. Other examples of this kind abound. Protons and neutrons are anomalous among the heavy particles in the sense that their optimum cross sections for inactivation (and hence their maximum RBE) occur at values of the mean free path that differs from the standard 2 nm. Belli et al. [1994] have specifically studied this effect experimentally but found it occurs at an LET value for protons of 37 keV/μm, which is somewhat lower than the 47 keV/μm deduced from the V-79 cell data used here. Also, Belli et al., comment that deuterons unexpectedly are somewhat less effective than protons at the same LET. However, this behavior is predictable at LETs near 47 keV/μm. By taking into account the respective penetration depths of the particles in the cellular material, the proton would have a mean free path for linear primary ionization averaging at ~2.7 nm compared with that of the deuteron of ~3.2 nm. The deuteron should therefore be slightly less effective than the proton for the same number of targets at risk, as is observed.

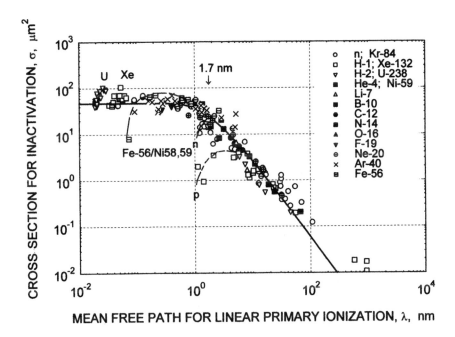

Figure 5.3. *Cross sections for inactivation of Chinese hamster V-79 cells are shown as a function of the mean free path for linear primary ionization. The curves displayed in this way permit deductions to be made on: target multiplicity, the physical reasons for the magnitude of the saturation cross section, and the cause of the Katz "tails" in the saturation region. [See* **Source of Data for Figures** *in References for Chapter 5.]*

Observed Saturation Cross Section for Biological Effect

Heavy Charged Particles

The observed saturation effect cross section, represented in Figure 5.3 by the horizontal line shown at small values of lambda, is about a factor of three times smaller than the geometrical cross-sectional area of the cell nucleus. Again, this is consistent with expectations if dsbs in the DNA are taken as the key targets at risk and allowance is made for the multiplicity of targets along the particle path. Thus, the known projected geometrical area of the DNA in the V-79 cell nucleus is 3 μm^2 (deduced from the mass of DNA and assuming spherical geometry) [Briden, 1988; Alkharam, 1997a] and the mean number of targets penetrated along a mean chord is ~12 to 15, determined again from the cell geometry and assuming that the content of DNA segments is uniformly distributed. The observed cross section for the radiosensitive region of the cell is thus expected to be $3 \times 12 = 36 \ \mu m^2$, which is in encouraging agreement with the experimental results shown in Figure 5.3 for V-79 cells. An independent check on the target multiplicity for heavy particle tracks can be deduced from published data on energy deposition in cylindrical volumes of DNA dimensions [Charlton et al., 1985], giving a value of 11 for alpha-particle irradiation [Watt, 1989]. More recently, Rydberg [1996] estimated a value of 14, which can be interpreted as the target multiplicity, deduced from the projected area of the chromatin.

X-Rays, Gamma Rays, and Electrons

The most striking features of the low-LET electron probe interpretation (Figure 5.1) are that there is no indication of the target structure that is manifest for the high-LET radiations and that the cumulative probabilities are more than an order of magnitude less than those for the heavy particles. However, as the reasoning in the following paragraph demonstrates, the results are indeed consistent with the conclusions reached for the heavy particles if the physical properties of the electron track are taken into account.

For the case of X- and gamma-ray external irradiations, the effect cross sections are the probabilities of inducing the biological endpoint per unit fluence of electrons in the equilibrium spectrum generated in the medium. For track-segment experiments with electron beams, the effect cross sections are probabilities of induction of the endpoint per unit fluence of incident electrons at the relevant instantaneous energy. It should not be necessary to distinguish between the two modes of experiment if the proper averaging is performed to obtain the mean free path, provided that the track segment experiment is a good descriptor of damage mechanisms. Electrons are most damaging when their mean free path for ionization is ~2 nm; i.e., at the very end of their tracks when their energy is near 200 eV and their range is only ~9 nm, which is orders of magnitude less than the dimensions of the cell nucleus. Therefore, only about one DNA segment is at risk along the track of a most damaging electron compared with more than 10 times that for the heavy particles that are more damaging because they have the capability to sustain the ~2 nm optimum spacing throughout the cell nucleus. (Think of this ratio of target multiplicity at risk ~1 to 10 as giving a quantitative physical explanation for the magnitudes of the quality factor for electrons with respect to heavy particles in radiological protection.) Electrons at energies below ~50 eV are innocuous. Fast electrons with energy larger than a few hundred electronvolts, having longer ranges, will place a larger number of targets at risk increasing to a maximum when their range equals or exceeds the cell nuclear dimensions ~7 μm. However, at these higher energies the consequential increase in the electron mean free path for ionization is accompanied by a steep lowering of the inactivation efficiency. It follows that the influence of the penetration ranges on the number of targets at risk must be taken into account in quantifying radiation quality. For the same reasons concerning this interplay between the damaging power of electrons, as determined by the linear mean free path for ionization, and the penetration range, *it is physically impossible for electrons to produce a saturation plateau in the effect cross sections*. The same applies to slow heavy particles of range less than the mean chord through the cell nucleus (see Figures 5.3 and 5.4). If the multiplicity of targets at risk is assessed in this way, then there seems to be no serious problem in determining the quality of ultrasoft X-rays or Auger electron emitters [Baverstock, 1988].

Note that in Figure 5.1 the effect cross sections used to determine the cumulative probabilities are expressed per heavy particle fluence and per electron fluence, respectively, for the high- and low-LET radiations. Thus the biological effect quantified in the high-LET ratio automatically includes the delta rays around the primary heavy particle tracks, whereas the latter ratio is due entirely to electrons in the relevant equilibrium electron spectrum. Consequently, the electron cross-section ratios shown in the plot of the electron and photon data are not collinear with the ions at the same value of the averaged mean free path, as might have been expected. This is because the heavy particles can sustain optimum damage throughout the cell nucleus, whereas the electrons cannot because of the many short-ranged electrons in the degraded spectrum. The fact that the electron curve is displaced downwards toward lower cross sections, and not upwards as may be expected for the damaging track-end radiations, reflects the very low intrinsic efficiencies of electrons when compared with heavy particles (which are discussed in the next section). (As an aside, it is obvious in nature that one should expect very low effectiveness for electrons, otherwise the human body could not cope so effectively with the half a million beta and gamma rays emitted every minute due to the naturally occurring radioactivity from the indigenous ^{14}C and ^{40}K isotopes.)

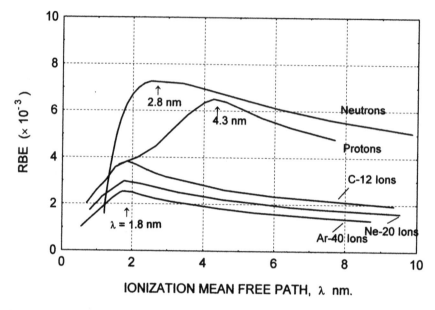

Figure 5.4. *Maximum RBEs normally occur at the position where lambda equals about 1.8 nm, as can be seen for C-12, Ne-20, and Ar-40 ions. However, the maxima RBE for protons and neutrons are displaced to larger lambda because the most damaging protons have ranges less than the cell diameter and there is a smaller number of targets at risk. Data are taken from Figure 5.1.*

The Role of Delta Rays Associated with Fast Heavy Ion Tracks

The Bethe formula for the collisional stopping power (i.e., LET) of fast ions (equation 4.23) is expressed as a function of z^2/β^2 and β^2, where β^2 is the reduced velocity squared and z^2 is the charge on the accelerated ion. z^2/β^2 can be interpreted in either of two ways. It is proportional to the yield of delta rays emitted per unit track and also to the source of the delta rays, the linear primary ionization. The linear primary ionization parameter should not be confused with the different quantity, the specific primary ionization used in Lea [1955], Belli et al. [1994], and Perris et al. [1986].

z^2/β^2 is theoretically invalid for slow ions. For fast particles, the reciprocal β^2/z^2 is directly proportional to the mean free path for linear primary ionization. The maximum spatial extent of the delta-ray penumbra generated by fast ions is proportional to β^2. It follows that if LET is a suitable parameter for description of radiation quality, then both the yield of delta rays ($\propto z^2/\beta^2$) and their spatial extent ($\propto \beta^2$) must be relevant at the track level. Similarly, if lambda, the mean free path for linear primary ionization in the charged-particle equilibrium spectrum, should transpire to be a good quality parameter, then it necessarily implies that the spatial distribution of the delta electrons must have negligible effect. As a test there are three methods by which the efficiency of damage due to delta rays may be evaluated.

1. The excess effect cross section, observed as tails (the Katz tails!) looping above the saturation value for the heaviest ions at low mean free paths in Figures 5.3 and 5.5, is attributable to the interaction of delta rays with cell neighbors of the primary target cell that enhances the cumulative probability of damage to greater than unity. In the case of argon ions in V-79 cells, the maximum cross section is observed at a mean free path, $\lambda = 0.4$ nm. The maximum cross section is ~70 μm^2, leaving an excess cross section of 34 μm^2 above the saturation value of 36 μm^2; i.e., 0.94 times the saturation value. Taking the cell diameter as d ~ 10 μm and the yield of delta rays as $1/\lambda$, then, in this rather crude calculation, the total number of delta rays accounting for the excess is $d*1000/\lambda = 2.5 \times 10^4$; i.e., the efficiency of damage per delta-ray track is ~$0.94/2.5 \times 10^4 = 3.6 \times 10^{-5}$! Similar arguments, if applied to the tiny *Bacillus subtilis* spores, indicate that the damage efficiency for delta rays from ^{40}Ar ions reaches almost 1%, presumably because of the much higher packing density of the spores, which places many more neighbor spores at risk. This is to be compared with the efficiency of 100% for a penetrating ion track having an interaction mean free path of ~2 nm.

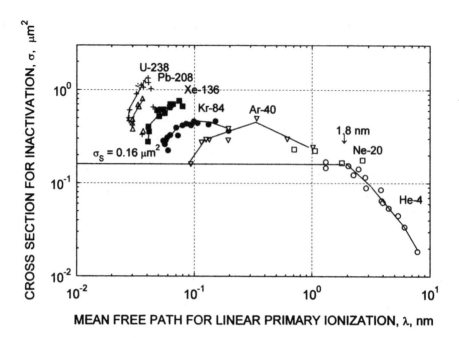

Figure 5.5. *Inactivation cross sections are shown for* Bacillus subtilis *spores irradiated with accelerated heavy particles.* [*Source of data is Baltschukat et al., 1986.*]

2. Fast ions of different types but with the same linear primary ionization will have the same yield of delta rays. As ions of different type have different effective charges, their β^2 will be different and hence the maximum extent of the spatial distribution of their delta-ray spectrum. If the delta-ray contribution to damage is negligible, then the effect cross sections for different particle types should be equal for a chosen value of lambda. The problem in testing is twofold. There is poor statistical precision of the cross-section data and there is an insufficient number of points for different ion types at the same lambda to make the analysis meaningful. The latter problem can be solved to some extent by selecting measured cross sections for various particle types having lambda values in a band between 6 and 22 nm; i.e., in the region where there are no complications with saturation effects. A scaling ratio based on the fitted expression $a(1 - \exp(-\lambda_{14}/\lambda)^b$ is used to interpolate the expected value of the cross sections at the midpoint reference value at 14 nm. If the delta rays have a negligible effect, then the scaled cross sections should be independent of the maximum energy of the delta rays. Within the wide spread of the results shown in Figure 5.6(a) for V-79 cells, there is no evidence for enhanced damage with increase in delta-ray energy for a specified type of particle. However, there is some indication that the cross section may increase with increasing Z of the

particles. Similarly, the results shown in Figure 5.6(b) for T1 human kidney cells indicate that the effect cross section at $\lambda = 14$ nm is a constant of 5 ± 2 μm^2 independently of Z for particle types greater than deuterons. There is little doubt that the precision could be much improved if survival experiments were conducted specifically to explore the point by using different ion types having lambda values in the nonsaturation region (see Figures 5.7 and 5.8).

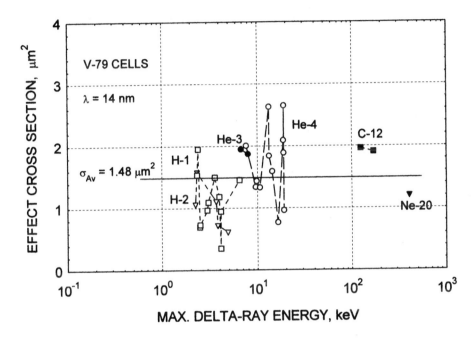

Figure 5.6(a). *Cross sections for inactivation of V-79 cells by different particle types having lambda values between 6 and 22 nm are normalized to 14 nm and plotted as a function of the maximum delta-ray energy to test if synergistic action by the delta rays contributes to the overall damage. Data are taken from Figure 5.3.*

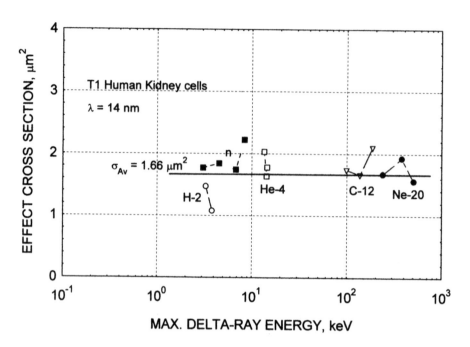

Figure 5.6(b). *Cross sections for inactivation of T1 human kidney cells by different particle types having lambda values between 6 and 22 nm and normalized to 14 nm are plotted as a function of the maximum delta-ray energy to test if synergistic action by the delta rays contributes to the overall damage. Data are taken from Figure 5.1.*

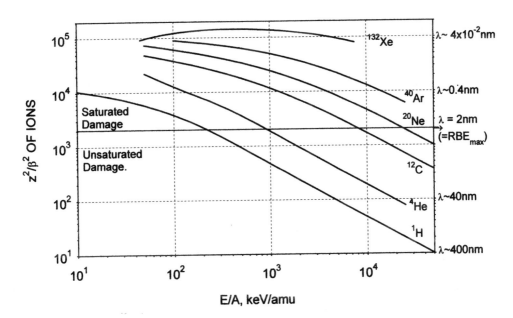

Figure 5.7. z^2/β^2 *is shown as a function of the specific energy, for accelerated ions in water, and its relationship to lambda. The maximum RBE should always occur at* $\lambda = 2$ *nm. At shorter values of lambda, saturation damage occurs and there will be no oxygen sensitivity. At larger lambda values, damage will be unsaturated, making this a more accurate region for the study of radiation effects.*

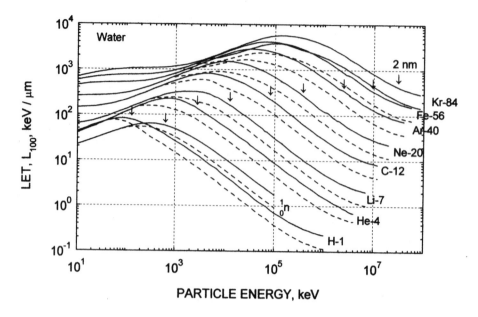

Figure 5.8. *The quality parameters LET and L_{100} are shown as a function of particle energy for the ions indicated. Optimum damage will occur, without saturation effects, at the λ = 2 nm position indicated by the vertical arrows. Note that the optimum damage occurs at positions well removed from the LET maxima, increasing with ion mass number. For this reason tests of radiation quality are best performed with heavy particles in nonsaturation conditions. The maximum effect for protons and neutrons is anomalous; see Figure 5.4.*

3. Simple geometrical evaluation leads to the expectation of very low damage efficiency for delta-ray irradiation of mammalian cells. As indicated previously, if a fast ion interacts with a cell nucleus to inactivate the nucleus with a cumulative probability of unity, then for the cumulative probability to exceed unity, the delta rays must have sufficient energy to reach neighboring cell nuclei and interact with their intrinsic DNA to produce lesions. To be able to penetrate to the adjoining nucleus, the delta rays must have an energy in excess of ~10 to 20 keV, depending upon whether the cells are plated or in suspension. The collection solid angle subtended at the point of emission of the delta ray will be something like 3% of 4π. Upon entering the neighbor-cell nucleus, to cause damage, the delta electron must intercept the DNA segments. The probability of this happening is ~10%, giving a net probability of 3×10^{-3}. Furthermore, the spacing of the ionizations must be correct to match that of the DNA if it is to produce a dsb; i.e., the electron track end must finish in the DNA, thereby reducing the chances

the chances by a further factor of 50 to 100 times because of the small fraction of eligible delta rays in the spectrum. (If the delta electron is sufficiently fast to penetrate the DNA, then the probability of interaction at the requisite 2 nm will be much smaller because of the mismatch with the DNA strand spacing.) The net result for the damage efficiency is of the right order (10^{-5}) to confirm the result obtained from the experimental data in (1.), mentioned previously. Further arguments in support of the proposed damage mechanisms are given in the next section.

Relevance of Absorbed Dose for Specification of Radiation Effects

Absorbed dose (more strictly, cema [Kellerer et al., 1992]) is the product of the equilibrium charged-particle fluence and LET. The LET is, in effect, a composite of the W value (mean energy per ion pair); the delta-ray yield per unit track length, given by the reciprocal mean free path for primary ionization, λ; and the average delta-ray energy, $T_{\delta,Av}$, i.e., $L = (W + T_{\delta,Av})/\lambda$. Since from the foregoing arguments it is the λ that best determines the quality of the radiation and not the LET, the average energy transferred ($W + T_{\delta,Av}$) in an interaction is not relevant. If one accepts the argument, based on experimental evidence, that the amount of energy transfer in excess of bond energies is irrelevant to the induction of radiation effects, then it is fundamentally wrong to use energy deposition as a quantifying parameter.

Other evidence that energy deposition is inappropriate can be found; for example, the occurrence of the wide variation with ion type of the range of values of the mean energy required to inactivate a molecule, as shown in Table 4.1. Two ionizations, if appropriately placed, are sufficient to break two single strands of the DNA whether the energy transfer is 10 eV or 1 MeV. All the evidence obtained here points to the conclusion that it is the *number* of events (dsbs in the DNA) caused by correlated pairs of ionizations that matter, not the energy transfer. Thus, two ionizations produced in the critical volume of a DNA segment need not induce a dsb. To do this the ionizations produced by the track must be correlated with the strands of DNA, like a template. If this is correct, then it would invalidate, on conceptual grounds, the use of volume quantities such as absorbed dose; the quality parameters ionization density and restricted LET; and the microdose quantities, lineal energy and specific energy density, because these quantities include the interactions of low-energy delta electrons.

The arguments presented previously and illustrated by Figures 5.5 and 5.6 suggest strongly that delta rays from heavy particles have negligible effect except in the saturation region at lambda values <<2 nm. Biophysically, an additional reason for their low capability of damage is that soft electrons of ~200 eV externally incident on the DNA have low probability of transition across the Z-disparate interface between the DNA and the intracellular fluid.

Delta electrons arising from heavy particle traversals through the DNA have mean free paths that are too large, >>2 nm, to supplement the primary track ionizations in the same DNA segment. Further support for this conclusion is that the magnitude of the RBE maximum, which occurs at $\lambda = 2$ nm for heavy particles, decreases with increasing atomic number of the particle, despite the enhanced range and energies of the delta rays, implying that the energy transported in the form of soft electrons does not contribute to the damage (Figure 5.4). The current series of experiments using charged-particle micro-beams to irradiate individual cells may eventually help to resolve the matter but, ideally, orders of magnitude improvement are required on the presently attainable beam resolution of about 1 micron spatial resolution [Reist et al., 1995; Folkard et al., 1995].

DNA Damage Is Due to Single-Track Action Except at Very High Doses (>10 Gy)

Chromosome Aberrations Are Due to Single Ion Tracks or Paired Electron Tracks

If the radiosensitive targets are taken to be segments of the DNA, then the notion of a quadratic component of action for production of dsbs becomes untenable except at very large doses. The argument is made as follows. The projected area of the DNA segment is approximately 2 nm by 3 nm. Assume that indirect action can occur due to the production of reactive species that will be effective in the surroundings of the DNA up to a diffusion distance of about 3 nm [Chatterjee and Holley, 1991]. The net geometrical area of the radiosensitive segment of DNA will be increased to approximately 70 nm². Assume also that an individual track can penetrate 15 DNA segments in traversing the mean chord through the cell nucleus, giving an effective geometrical cross-sectional area of ~10^3 nm². If mammalian cells are irradiated with, for example, 250 kV X-rays, a dose of 8 Gy reduces the surviving fraction well into the so-called quadratic region of the dose-response curve. This dose is equivalent to an equilibrium electron fluence, including the secondary electron cascades, of about 10^9 per cm^2 with a track average LET of 4.3 keV/μm (refer to equation 5.3).

$$\Phi(cm^{-2}) = \frac{6.25 \times 10^8 \times \rho(g/cm^3) \times D(Gy)}{\overline{L_T}(keV/\mu m)} . \qquad (5.3)$$

The mean number of tracks penetrating the DNA is equal to 10^9 cm^{-2} × 10^3 nm^2 = 0.01! This assumes that the electron tracks are 100% effective at producing dsbs, which, of course, is a gross overestimate. However, Lea [1955], Chadwick and Leenhouts [1981], and Harder [1986] have proposed that cell inactivation is a consequence of pairwise lesions in chromatids to produce chromosome aberrations. The chromatids have geometric cross-sectional areas of approximately 1 μm^2; i.e., 10^3 times larger than the geometrical sensitive area of a DNA segment. Therefore, for pairwise lesions in chromosomes undergoing electron irradiation, almost all the radiation action is expected to be quadratic.

For heavy particle irradiations, analogous arguments show that DNA damage is essentially a single-track process. Similarly for production of chromosome aberrations, pairwise lesions are predominantly a single-track effect because the heavy particle can sustain its maximum damaging power throughout the whole cell nucleus, whereas the electron has much less chance, by orders of magnitude, of achieving this. These conclusions are consistent with the observed factor of 4 reduction in the cross section for production of chromosome aberrations by heavy particles [Watt, 1989] when compared with those for cell inactivation at the same particle mean free path for linear primary ionization [Alkharam, 1997a]. Edwards [1994] has demonstrated that the yields of chromosome aberrations have a linear dependence with absorbed dose for heavy particles as would be expected for a single-track effect. As an average of four chromosome aberrations are required to account for cell inactivation, it follows that the chromosome aberration alone is not the prerequisite criterium for cell death. The results of the analysis made here suggest that the dsb in the DNA is in itself sufficient to inactivate, which is still consistent with the proposal that two pairs of dsbs are required for the chromosome aberration.

Comment on "Low Dose"

The definition of "low dose" has caused some controversy in conventional dosimetry because of the need to partition the linear and quadratic responses in the survival curves—essentially to deal with dose-rate effects [e.g., Hall, 1994]. In a fluence-based system, low dose can be unambiguously interpreted as the induction of the bioeffect, due to a single track, which occurs when the mean number of events $\sigma_{B,sat} \times \Phi_{eq} < 1$. $\sigma_{B,sat}$ is the saturation cross section for induction of the cellular effect by charged particles in the relevant equilibrium fluence, Φ_{eq}. For inactivation of reproductive capacity, the corresponding fluence is less than $10^8/\sigma_{B,i} \sim 10^8/50$ cm^{-2} for any type of heavy charged particle and $\sim 10^8/4$ cm^{-2} for electrons. If transposed into the absorbed dose system, these fluences are equivalent to low doses of $1.6 \times 10^{-3} \times L_T$ (keV/μm) Gy and $2.5 \times 10^{-2} \times L_T$ (keV/μm) Gy, respectively. (Values of L_T for equilibrium spectra are tabulated in Watt, 1996.) As an example, for 250 kVp X-rays and 5.5 MeV alpha particles, the defined low dose is in the range 100 to 250 mGy, which could be considered as a lower limit for onset of the bioeffect in the absence of repair. Scaling to the other endpoints of chromosome aberrations, oncogenic trans-formations and DNA dsbs can be performed on the basis of the saturation effect cross sections that occur at a mean free path for ionization of 2 nm [Alkharam, 1997b].

Relative Biological Effectiveness: Maximum RBE

As noted in Chapter 3, the concept of RBE is meaningless at low doses when not all the sensitive volumes receive at least one hit. For this reason, it is better simply to use the measured effect cross section because this is an *absolute* measurement of radiation effectiveness, making the use of ratios unnecessary.

There are other general limitations to RBE. Only a tiny fraction of the energy expended (<<1% for heavy particles) goes into the production of lesions. The corresponding ratio of useful signal dose to unwanted background dose (Table 4.1) means that the statistical error in the RBE must be very large indeed, making it insensitive to important detail. For example, the excess damage produced in V-79 cells or *Bacillus subtilis* spores irradiated by argon ions with lambda values of 0.4 nm, shown in Figures 5.3 and 5.5, respectively, makes a large contribution to the effect cross section [Kraft, 1987], yet it is almost indistinguishable in the magnitude of the RBE. Cross sections for biological effects are related to the RBE through the LET ratio in the equation:

$$RBE = \frac{D_{\gamma,37}}{D_{ion,37}} = \frac{\sigma_{ion}}{\sigma_\gamma} \times \frac{L_\gamma}{L_{ion}} \tag{5.4}$$

where the symbols have their usual meaning.

Often RBEs are plotted as a function of LET to explore the occurrence of maximum RBE. This is not a valid procedure for heavy particle radiations of different types because each radiation has a unique RBE/LET curve with a single maximum [Figure 5.9(c)] that occurs at different values of LET, according to particle type. However, if the RBEs are plotted against the relevant value of the mean free path, the maxima for mammalian cells always occur at the common 2.0 ± 0.3 nm (~ 2 nm) position. This indicates that the maximum RBE is caused not by a saturation effect associated with energy wastage, but by a template effect that is not too dissimilar from the resonance concepts hypothesized in different forms by Yamaguchi and Waker [1982] and Watt et al., [1984], although it is not strictly a resonance per se. The biological lesion is produced when the randomly created ionizations in the radiation track have the requisite 2-nm spacing and each ionization occurs in a strand of the DNA segment simultaneously matching exactly to produce a dsb. As the mean free path decreases below ~2 nm, the onset of saturation indeed occurs, but this is because of the rapid decrease in ionization mean free path as the radiation tracks slow down. The results are not yet accurate enough to prove that the effect goes through a maximum at the critical mean free path before the onset of true saturation; examination of some of the data sets do suggest a maximum; i.e., the ionizations must be separated by the intrahelix distance of ~2 nm rather than occur anywhere within the volume in the 2-nm critical length.

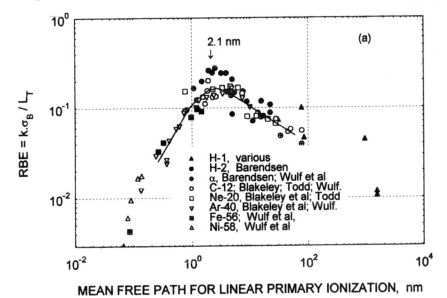

Figure 5.9(a). *RBE for inactivation of mammalian cells is shown as a function of the quality parameter λ for different heavy particle types.*

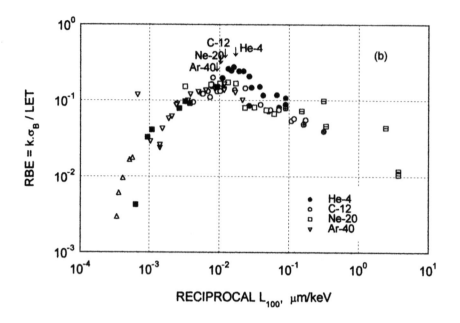

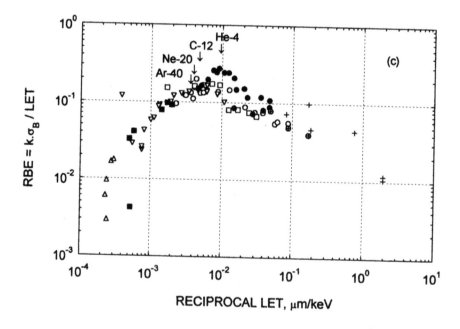

Figure 5.9(b, c). *RBE for inactivation of mammalian cells is shown as a function of the quality parameters $1/L_{100}$ and $1/L_\infty$, respectively, for different heavy particle types.*

These findings are consistent with and explain the results reported by Cox et al., [1977] and Thacker et al., [1979] that RBE reaches a maximum value for some fast heavy particle types at the same value of z^2/β^2 [Watt et al., 1985]. As indicated earlier, protons and neutrons are exceptions to this rule because, when their mean free path is at the optimum value, their relevant ranges are less than the cell diameter, the target multiplicity at risk reduces, and hence the damaging efficiency falls. Consequently, for these radiations the maximum RBE is shifted to lower LET, or longer mean free path for ionization, as shown in Figures 5.4 and 5.9.

Additionally, careful examination of the neutron data shown in Figure 5.3 reveals another manifestation of the reduced multiplicity of targets at risk on account of the short ranges of some of the protons in the recoil spectrum. This causes an overall reduction in the neutron effect cross section when compared with the heavy charged-particle radiations, represented by helium and carbon ions, as illustrated in Figure 5.10.

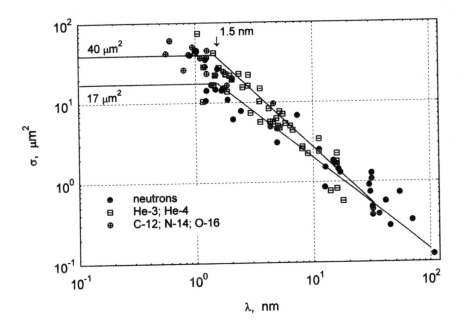

Figure 5.10. *When compared with heavy charged particles (He, C, N, and O ions) the effect cross sections for neutrons are seen to be appreciably smaller on account of the proportion of short-range protons in the equilibrium spectrum. Consequently, neutrons are always less damaging than heavy charged particles at the same lambda and equal charged-particle fluence.*

In the thesis presented here, heavy particles with the same lambda produce the same effect, whereas if effectiveness is measured in terms of RBE, anomalies arise simply because of the invalid role of the LET. For example, Folkard et al., [1995] find that p and d irradiations are more effective than alpha particles of equivalent LET by comparing the RBE at 10% survival of V-79 cells. This observed trend in quality is predictable from the mean free paths for linear primary ionization.

System of Dosimetry

Consideration of the foregoing analyses of the influence of track structure properties of radiations on the bioeffectiveness leads to new conclusions on the fundamental mechanisms involved. This, in turn, points the way to modeling the process quantitatively and hence to establishing a better, unified system of dosimetry, including changes to the basic requirements for instrumentation needed for its practical implementation.

The conclusion to be drawn is that the basic mechanism of radiation damage to normal mammalian cells is the correlation of two ionizations, which are spaced at about 1.5 to 2 nm along single-particle tracks in the relevant charged-particle spectrum, with the similarly spaced strands of the intranuclear DNA. For indirectly ionizing radiations, the relevant charged-particle spectrum is the equilibrium spectrum. A template action is proposed because it is most consistent with the observations. Ionizations produced at other locations or at other orientations do not fulfill this requirement and so energy deposition cannot apply on both fundamental and practical grounds. Conceptual arguments against the use of conventional microdose quantities and restricted LET suggest that these parameters too are unsuitable, even at the nanometer level, although it may be possible to derive approximate adjustments using appropriate probability functions [Kellerer and Rossi, 1978] that may be of sufficiently good approximation to be of application in practice. The biological effects discussed are found to depend on the *number* of such paired interaction events, which are independent of the energy transfer. No clear evidence is found for dual radiation action in inactivation of cells. If present, dual radiation action would be observable only at very high doses, even for low-LET irradiations, as discussed in Chapter 6.

As the cumulative probability of damage is an absolute measure of the biological effectiveness for the irradiation conditions used, it follows that a system of dosimetry can be more rigorously based on the equilibrium charged-particle fluence and the effect cross-section ratio to provide a unified system applicable to any radiation type, whether delivered externally or internally.

Radiation quality appears to be more effectively determined by lambda, the mean free path for linear primary ionization, for the charged-particle tracks in the relevant equilibrium spectrum. Lambda is a linear property of the radiation track unlike specific ionization density, or LET and its derivatives, which represents spatial (volume) properties of the tracks.

Modeling the proposed mechanism is of value in several specific ways. It can indicate methods of testing the conclusions reached; provide a scientific framework for specifying instrument response functions and design; resolve the problem of extrapolating effects to low doses near environmental levels; and indicate the requirements for validating any proposed new system of risk control in radiation protection. From the conclusions reached here the practical system could be a unified one, independent of radiation type, and possibly of biological endpoint by application of scaling relationships. Chapter 6 discusses a model of radiation bioeffectiveness and its application in a fluence-based system.

6.

Model of Radiation Bioeffectiveness and Its Application in a Fluence-based System

Calculation of Biological Effectiveness in Mammalian Cells

The conclusions reached in Chapter 5 on the biophysical mechanisms of radiation action can be easily accommodated in a radiation effect simulation model and tested against experimental data taken from the literature. However, as the induced damage is concluded to be a single-track effect, dose-rate effects cannot be relevant, except at very high doses, and so another mechanism must be found to account for the shoulder frequently observed in the logarithmic survival response in mammalian cells and attributed to enzymatic repair of damage [Hall, 1994]. Few models have dynamic terms that attempt to account for the effects of repair; nevertheless, some unification of data has been achieved by Curtis [1986] using the dual concept of lethal, potentially lethal (LPL) lesion. However, knowledge is still insufficient on the detailed dynamics of the repair mechanism and on how best it may be quantified [Steel, 1989].

The hypothesis supported here is that it is the duration of the irradiation rather than the dose rate that determines the shape of the survival curve. There is a rather subtle difference between the two because the modifying factors associated with the duration of the irradiation and the dose rate are often interrelated. In practice, two experimental situations may arise: (1) Increased doses can be delivered at a fixed dose rate by changing the duration of the irradiation. This is expected to cause a shoulder on the observed survival curve; (2) Alternatively, different doses may be delivered in a fixed irradiation time

87

interval by changing the dose rate. In this latter circumstance, for single-track action, no shoulder is expected in the survival curve. In both cases it is surmised that the duration of the radiation exposure is the fundamental factor, not the dose rate. Thus in deriving a model for cell survival, the additional factors generally considered to govern the shape and magnitude of the survival curves must be taken into account, viz., the duration of the irradiation, whether the cell population is synchronized or asynchronized, and the changing radiosensitivity in progression through the cell cycle. The relevant "unrepair functions," $U(Z,t_{m,r,c,i})$, which cope with the temporal aspects of the irradiation, are expressed in equations 6.1 and 6.2.

Repair of Damage, the $U(Z,t_{m,r,c,i})$ Function

The function $U(Z,t_{m,r,c,i})$ represents the probability that radiation-induced dsbs in the DNA will remain *unrepaired* at the end of the cell cycle. The derived expressions for synchronized and asynchronized cells are based on the assumption that there exists a mean rate of repair of lesions, and follows a logic that is similar to that used by Lea [1955], Harder and Virsik-Peuckert [1984], and Kiefer [1987] but with critical modification to allow for a specified damage-fixation time, the variation in radiosensitivity during the individual cell cycle, and synchronization of the cell population through the mitotic cycle.

For Synchronized Cells

$$U(Z,t_{m,r,c,i}) = \frac{1}{t_i} \int_{t_c}^{t_c+t_i} Z(F,t) \cdot e^{-(t_m-t)/t_r} \cdot dt \qquad (6.1)$$

For Asynchronized Cells

$$U(Z,t_{m,r,i}) = \frac{1}{t_i} \cdot \frac{1}{t_m} \int_0^{t_m} \int_{t_c}^{t_c+t_i} Z(F,t_j) \cdot e^{-(t_m-t_j)/t_r} \cdot dt_j \cdot dt_c \qquad (6.2)$$

$Z(F,t)$ is an empirical function that makes a rather crude approximation for the change in radiosensitivity during the cell cycle [Sinclair, 1972; Hall, 1994] by using a constant radiosensitivity in the G1 phase, a triangular response that minimizes in the less sensitive S phase, and followed by increasing sensitivity until mitosis. t_i and t_m are, respectively, the duration of the irradiation and the time, taken to be equivalent to the cell cycle, at which damage is considered fixed (at mitosis). t_r is the mean time (1 - 1/e) taken to repair 63% of the lesions. t_c, which applies only to synchronized cell populations, is the time into the cell cycle at which the irradiation begins.

The "unrepair functions" occur in the exponent of the equation for the surviving fraction (e.g., equation 6.4) and have the novel properties of producing a variety of shapes. For asynchronous normal cells [Figure 6.1(a)], the slopes are predicted to be always shoulderless in the limiting cases of acute irradiation ($t_i \ll t_m$) and of chronic irradiation ($t_i \sim t_m$), with shouldered or shoulderless survival between these limits, depending on the mean rate of repair. Shorter repair times lead toward shoulderless survival [Figure 6.1(a)]. At equal doses, chronic irradiation is always less damaging than acute irradiation. If the cells are synchronized [Figure 6.1(b)], the situation is more complex. Acute irradiations lead to shoulderless survival curves. As the duration of the irradiation increases, the curve tends to be pure exponential until the start of the irradiation is moved several hours into the cell cycle, at which stage a shoulder may develop that becomes progressively more pronounced. Manifestation of these predictions is controlled by the effect of the $Z(F,t)$ function for the radiosensitivity in the cell cycle.

Radiosensitivity

As the radiosensitivity changes with the position in the cell cycle, and is dependent on the cell type, it is challenging to describe its underlying cause. The critical radiosensitive lesions are considered here to be dsbs in the DNA that are produced by single ionizing tracks traversing a mean chord through the cell nucleus. On this hypothesis, the average radiosensitivity of a cell is determined by the mean number of DNA double-stranded segments at risk per unit fluence, represented by $\sigma_{g,DNA} \cdot n_0$ in equation 6.3. However, the cell is a dynamic entity in which the DNA packing changes throughout the cell cycle, reaching a maximum near mitosis and minimizing in the S phase. Therefore it seems likely that the $Z(F,t)$ function for the radiosensitivity in equation 6.1 can be interpreted as the changing packing ratio of the DNA with respect to the mean value. Consequently, the $U(Z,t)$ function represents a modifier to the mean number of DNA segments at risk per unit fluence, given by $U(Z,t) \cdot \sigma_{g,DNA} \cdot n_0$, which corrects for the influence of the DNA dynamics, and of the repair probability for lesions, on the cell survival, thereby constituting the inherent radiosensitivity of the cell under the prevailing radiation conditions. These are cellular properties. The *degree* of radiation damage will depend on the radiation quality; i.e., on the physical factors governing the efficiency with which the radiations can activate the radiosensitive sites.

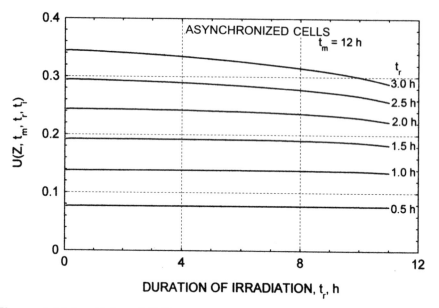

Figure 6.1(a). *Values of $U(Z,t_m,t_i)$, the probability that dsbs induced in DNA will remain unrepaired by mitosis, for asynchronized cells are shown as a function of the irradiation time for various mean repair times, t_r.*

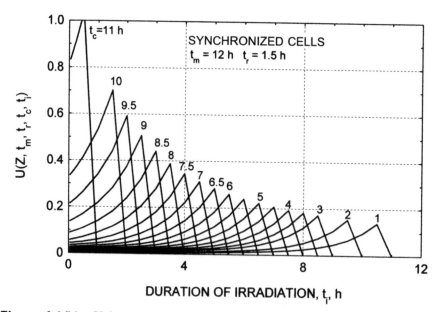

Figure 6.1(b). *Values of the function $U(Z,t_m,t_o,t_i)$, which represents the probability that dsbs in DNA will remain unrepaired at mitosis ($t_m = 12$ h), are shown as a function of the irradiation time for synchronized cells having a mean repair time of 1.5 hours. The times, t_o, indicated are the times into the cell cycle at which irradiation begins.*

Absolute Bioeffectiveness of Ionizing Radiation: Number of dsbs Induced in the Intracellular DNA

For any type of mammalian cell, the probability of damage per unit fluence for single-track inactivation can be expressed in terms of the proposed mechanisms of damage, including direct and indirect action, as an effect cross section, σ_B. The product of σ_B and the equilibrium fluence of charged particles, Φ_{eq}, defines the absolute bioeffectiveness, B, of the radiation field. At the most fundamental level, B also represents the mean number per cell of unrepaired dsbs produced in the intranuclear DNA by the relevant charged-particle radiation. These factors can be combined into the well-known Poisson relation to obtain the survival fraction of irradiated cells, F, given by, $F = \exp(-B)$. The bioeffectiveness can be expressed directly in terms of fundamental physical and biophysical quantities (defined in Table 6.1) as:

$$B = \sigma_B \cdot \Phi_{eq}$$

$$\sigma_B = \sigma_{g,DNA} \cdot \frac{n_0}{\overline{d}} \cdot \left[\int_E R(E) \cdot \Phi(E)_{eq} \cdot \varepsilon_{dsb}(\lambda(E)) \cdot dE \right] \cdot U(Z, t_i) \qquad (6.3a)$$

By taking fluence-weighted averages for R $(<d)$ and λ in the damage efficiencies, the effect cross section in equation 6.3a simplifies to equation 6.3b, which is adequately accurate for most applications (see Figure 6.2). n_0 is taken as 15 for densely ionizing radiation and as unity for sparsely ionizing radiations in equation 6.3b.

$$\sigma_B = \sigma_S \cdot \varepsilon_{dsb}(\overline{\lambda}) \cdot U(Z, t_i)$$

$$\sigma_S = \left[\sigma_{g,DNA} \cdot n_0 \cdot \frac{\overline{R}}{\overline{d}} \right]; \quad \frac{\overline{R}}{\overline{d}} \le 1 \qquad (6.3b)$$

Dual-track action may occur at exceptionally high dose rates, in which case the effect cross section can be written as the cumulative probability, defined as the ratio to the cross section for saturation damage:

$$\frac{\sigma_B}{\sigma_S} = \left[\varepsilon_{dsb}(\overline{\lambda}) + 2 \cdot \frac{\left(\varepsilon_{ssb}(\overline{\lambda}) \right)^2}{n_0^2} \cdot \sigma_S \cdot \Phi_{eq} \right] \cdot U(Z, t_i)$$

$$= \varepsilon_{dsb}(\overline{\lambda}) \cdot \left[1 + \frac{2}{n_0^2} \cdot \sigma_S \cdot \Phi_{eq} \right] \cdot U(Z, t_i) \qquad (6.3c)$$

Equation 6.3c indicates that dose-rate effects cannot be greater than 1% at 37% survival; i.e., $\sigma_B \cdot \Phi_{eq} = 1$.

Table 6.1. *Definition of quantities used in equation 6.3.*

Quantity	Definition
σ_B	The cross section for the stated biological endpoint, in this case inactivation of mammalian cells. It is also the total cross section for induction of dsbs in the DNA of the cell nucleus.
σ_S	The saturation effect cross section, shown by the horizontal line drawn to smaller values of λ through the point at $\lambda = 1.8$ nm (see Figures 5.1, 5.5). σ_S is spectrum-dependent for tracks having $R/d \leq 1$. For heavy ions in V-79 cells, $\sigma_S \sim 45$ μm^2, and for equilibrium electron tracks, $\sigma_S \sim 3$ μm^2, as shown in Figure 6.2. σ_S, the saturation cross section defined previously, can be expressed as $\sigma_{g, DNA} \cdot n_0 \cdot R_p/d$. If $R_p > d$, $R_p/d = 1$. The cell radiosensitivity function, $Z(F, t_i)$, is presumed to correct for changes in the DNA configuration during the cell cycle.
σ_B / σ_S	This ratio represents the cumulative probability for inactivation. It can have a value greater than unity in the saturation region for heavy ions where delta-ray effects may contribute to the damage, as shown in Figures 5.3 and 5.5.
d	The mean chord length through the cell nucleus (~ 7 μm).
R_p	The mean projected range of the relevant tracks. If $R_p > d$, $R_p/d = 1$, which partially allows for the reduced multiplicity of targets (number of DNA segments) at risk for the short-range "stopper and insider" tracks in the cell nucleus.
n_0	The number of dsb segments at risk per track traversal (~ 15 for fast ions). n_0 will depend on the dimension of the mean chord distance through the cell; i.e., whether it is flattened by plating or free in suspension.
$\sigma_{g, DNA}$	The projected geometrical cross-sectional area of the intranuclear DNA. It is dependent on cell type and on the stage in the cell cycle. For V-79 hamster cells, the projected area of the DNA averaged over the cell cycle is about 3.0 μm^2.
Φ_{eq}	The equilibrium fluence of the relevant charged particles in the cell nucleus. Values for most radiations are given in Watt, 1996.

Quantity	Definition
$\Phi(E)_{eq}$	The differential fluence in the energy spectrum.
$\varepsilon_{dsb}(\lambda)$	The efficiency for production of dsbs in the DNA by direct plus indirect action, given by: $\varepsilon_{dsb}(\lambda) = (1 - e^{-(2\Lambda + s)/\lambda})^2$; $\lambda_0 = 1.8 \pm 0.3$ nm.
λ	The mean free path for linear primary ionization of the equilibrium particle tracks [Watt, 1996]. (More strictly, the mean *spacing* between ionizations rather than the mean free path is the relevant quantity, especially for electron tracks [Grosswendt, 1997], but numerical data on spacings are not yet available.)
$\varepsilon_{ssb}(\lambda)$	$= 1 - e^{-(2\Lambda + s)/\lambda}$. The efficiency for ssb production in the DNA by direct plus indirect action in a single DNA strand of width s nm.
Λ	(~ 1.5 nm) is the mean diffusion length in the immediate vicinity of the DNA molecule for active radicals (indirect action) originating from a line source. Radiosensitization due, for example, to the presence of oxygen will occur only if the λ value for the radiation is greater than $\lambda = 2$ nm; i.e., in the region of unsaturated damage. Radiosensitization or radioprotection due to chemical additives is accommodated by the resulting change in the diffusion length, Λ, in the efficiency term, $\varepsilon_{dsb}(\lambda) = (1 - e^{-(2\Lambda + s)/\lambda})^2$, of equation 6.3.
s	~1 nm. The thickness of a single strand of DNA.
$U(Z, t_{m,r,i})$	The probability that dsbs in the DNA remain unrepaired.
t_m	The time to mitosis.
t_r	The mean repair time.
t_i	The duration of the irradiation.

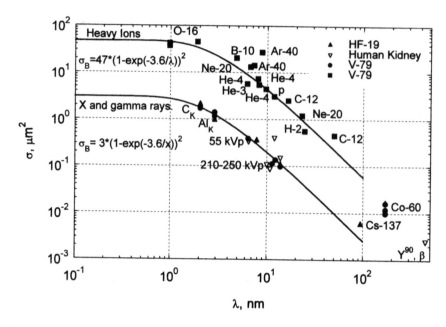

Figure 6.2. *Calculated cross sections (equation 6.3b) as a function of lambda, the mean free path for primary ionization, are compared with some experimental values, mainly for V-79 cells. X- and gamma-ray data for HF-19 cells have been normalized at 2 nm.*

Total Effect Cross Sections in the Saturation Region

Effect cross sections calculated from equation 6.3a or 6.3b are intended to apply to the unsaturated region of damage where $\lambda > 2.0$ nm; i.e., to the region where the delta-ray damage contribution from heavy particle tracks is negligible. In the saturation region, the yield of delta rays from heavy tracks in the vicinity of a cell nucleus can be enormous as the yield is proportional to $1/\lambda_i$ (Figure 5.3). An approximate correction factor, which gives the maximum effect cross section of the delta-ray "tails" as $\sigma_S \cdot (1 + 2 \cdot 10^{-5} \cdot R_{el}/\lambda_i)$, can be easily applied. R_{el} is the projected range of the maximum energy delta ray at an ion mean free path λ_i and σ_S is the saturation cross section for the heavy particle. If necessary, more accurate accounting for the delta-ray spectrum can be simply performed by analytical or Monte Carlo methods.

A Unified Model for Damage to Mammalian Cells by Action of Single Charged-Particle Tracks

As indicated previously, the survival fraction for the proposed single-track action can be expressed as:

$$F = \exp\left[-\sigma_B(\lambda, U(Z,t)) \cdot \Phi_{eq}\right] \tag{6.4}$$

in which Φ_{eq} is the equilibrium fluence of charged particles generated by the incident radiation field. For indirectly ionizing radiations, the equilibrium fluence is conveniently obtained from the track definition of fluence, which equates the continuous slowing down approximation (csda) range to the equilibrium fluence per unit source concentration of tracks [Chilton, 1979]. Thus,

$$F = \exp\left[-\sigma_B \cdot \sum_j \left(R_j \cdot N_{T,j} \cdot \sigma_{inc} \cdot \Phi_i\right)\right] \tag{6.5}$$

The summation, carried out over the source concentration, $N_{T,j} \cdot \sigma_{I,j} \cdot \Phi_i$, of charged particles type j having ranges R_j in the csda, represents the particle fluence generated. (Equation 6.6, which applies to the electron fluence from radionuclides incorporated into mammalian cell structure, is seen to be of similar form to the fluence term in equation 6.5.) Because the exponent in all of the survival equations represents the mean number of dsbs induced in the DNA of mammalian cells, the technique provides, in principle, a method of absolute dosimetry that is independent of radiation type and applicable to internal or external exposures.

Furthermore, there is supporting evidence that scaling factors may be applied, using equation 6.3b, to the inactivation data to determine the cumulative probabilities for the induction of other biological endpoints such as chromosome aberrations, oncogenic transformations, and mutations [Alkharam, 1997], which offers the possibility of encompassing the whole spectrum of DNA-dependent bioeffects into a unified system for protection purposes. The need for artificial quality factors or radiation weighting factors does not arise.

Practical implementation of the unified system requires that instrumentation be designed to have a radiation response which simulates that of the DNA in the cell nucleus [Watt and Alkharam, 1995; McDougall et al., 1997].

Model Prediction of Effect Cross Sections for V-79 Cells

To test the validity of the single-track damage mechanism, a comparison is made, in Figure 6.2, of the absolute bioeffect cross sections calculated from equation 6.3b with experimental measurements taken from the literature. The experimental cross sections were determined, for V-79 Chinese hamster cells, from the initial slopes of survival curves measured during the past 20 years at several different laboratories, randomly selected from the extensive database compiled by Alkharam [Alkharam and Watt, 1997]. The initial slopes were selected to avoid accounting for repair. Taking the projected cross-sectional area of the DNA for V-79 cells as 3 μm^2, and the diffusion length, $\Lambda = 1.3$ nm, for indirect action, the calculated cross section for any heavy particle is given by $\sigma_B = 47 \cdot [1 - \exp(-3.6 \lambda)]^2$ μm^2 and for X- and gamma rays as $\sigma_B = 3 \cdot [1 - \exp(-3.6 \lambda)]^2$ μm^2, valid for $\lambda > 2$ nm. The effect cross section is also the mean number of initial DNA dsbs per cell per unit charged-particle fluence.

Despite the encouraging agreement between the calculated and the experimental values that have a rather large spread of errors, there is a serious anomaly for the ^{60}Co data. The cross sections are much too large and do not appear to have a logical trend. Eight independent measurements with ^{60}Co are in agreement within the typical errors (~30%) that the cross section is about 0.0136 ± 0.004 μm^2 determined for dose rates ranging from 1.2 to 16 Gy/min. The magnitudes and trends of the individual results do not indicate any dose-rate dependence notwithstanding the standard errors. Unfortunately, there does not appear to be any measurements of cross sections for inactivation of V-79 cells by ^{137}Cs or other monoenergetic gamma rays near the ^{60}Co energies, but a comparison can be made if results for T1 human kidney cells and HF-19 cells are normalized to the V-79 results near the 2-nm position. Then the results for ^{137}Cs gamma rays and ^{90}Y beta rays confirm the apparent anomaly with [60]Co. Apart from some differences caused by the averaging of data, a plausible explanation for the anomaly is that the damage is enhanced by the simultaneous emission in the ^{60}Co decay of two gamma rays (1.17 and 1.33 MeV) that are angularly correlated and that will give rise to separate but time-coincident electron cascades in the equilibrium spectrum. Because the time separation between the cascades is negligibly small, the radiation will act as if it were delivered synchronously. Simple calculation based on the probabilities of induction of correlated ssbs shows that the observed effect cross section would be enhanced by a factor of 4 times compared with radiation resulting from independent gamma-ray cascades. If the cross section is divided by four, the anomaly disappears.

Dosimetry of Incorporated Radionuclides

Electron-emitting radionuclides, especially those decaying by Auger electron emission, are known to be excessively damaging when incorporated into the nuclear component of mammalian cells [Howell and Rao, 1996]. Indeed, the claim is made that Auger electron emitters, such as ^{125}I, can attain the damage level of high-LET particles. The subject is of considerable topical interest because of the importance in nuclear medicine as well as the general implications for radiation protection. Conventional dosimetry is found to be inadequate to quantify the observed bioeffects [Humm et al., 1994], thereby presenting a challenge to the perceived mechanisms of radiation action and to modeling.

Various validation tests have been conducted during the development of the model that has as key elements the notion of predominantly single-track action (therefore dose rate independent) and a repair mechanism based on the duration of the irradiation and on the residual time available for repair (refer to equations 6.1, 6.2, and 6.3). With regard to residual time available for repair, the cell survival experiments of Hofer et al., [1992] and Hofer and Bao [1995] are of special interest as they are unique in the manner by which the exposure times are controlled to specific points in the time cycle of synchronized cells. The experiments were performed at constant dose rate using Chinese hamster ovary (CHO) cells labeled with ^{125}I-iodo-deoxyuridine. By accumulating the decays at liquid nitrogen temperatures, the initial radiation damage can be, in effect, constrained to a known point in the cell cycle. Equation 6.2 applies for estimation of the unrepaired fraction of dsbs in the DNA.

Although the model given by equation 6.3 is evolved from fundamental principles comprising well-established physical quantities, there are two variable parameters of magnitudes that are not specifically known with certainty—the mean time taken for the cell's repair of DNA dsbs and the radical diffusion length in the presence of natural scavengers. The model predicts that for doses of electrons less than several kilogray, the radiation action should give rise to pure exponential survival in the Hofer-type experiments. However, in the special circumstances of an Auger electron emitter, such as ^{125}I, which may emit simultaneously an average of seven highly localized electrons per decay [Charlton and Booz, 1987; Weber et al., 1989], there may be the possibility of a very small increase in the number of coincident events due to the presence of the multiple electron tracks in the immediate vicinity of the DNA. The effectiveness of the electrons remains small, even if the isotope is tagged to the

DNA molecule, because of the unfavorable linear primary ionization, the geometry of emission, and the inherent protection given by the change in electron density across intracellular interfaces.

Knowledge of the electron fluence is necessary if equations 6.1 to 6.3 are to be applied. In the case of a distributed radionuclide, the fluence is best deduced from the track (range) definition [Chilton, 1979]. Use of the csda range also provides a convenient means of automatically allowing for the reduced multiplicity of cellular targets (DNA segments) at risk, which immediately solves one of the inadequacies of conventional dosimetry. Thus the electron fluence in the target medium is given by the product of the concentration of radionuclide and the csda range; i.e.,

$$\Phi_{eq} = \frac{A \cdot t_i \cdot n_{el}}{V} \cdot \sum_j [f_j \cdot R_j] \qquad (6.6)$$

where A is the total activity (number of decays per second) per cell nucleus of the incorporated radionuclide; t_i is the duration of the irradiation; n_{el} is the total number of electrons emitted per decay; V is the volume of the cell nucleus; and f_j is the fraction of electrons in group j per decay having csda range, R_j. In the case of ^{125}I, the decay was divided into four main electron groups, j comprising the K and L conversion electrons and the K, L, and M Auger electron groups yielding a mean of 5.7 electrons per decay [Weber et al., 1989; Watt, 1996]. The survival fractions, shown in Figure 6.3(a), and the mean number of unrepaired dsbs in the intranuclear DNA, shown in Figure 6.3(b), were calculated directly from equations 6.1 to 6.3(a) and by obtaining a simultaneous least squares fit to the 50 experimental data points comprising the five survival curves. If the projected area of the DNA segments at risk, $\sigma_{g,DNA} \cdot n_0$, for the CHO cells is taken to be 56 μm^2, the mean diffusion length for radical action is found to be 2.7 nm and the average time constant for repair of the dsbs is 2.05 hours.

The general trends of survival with irradiation time and cycle time are seen to be reasonably well described in Figure 6.3(a), as is the Λ-dependent effect of adding a radioprotector, which makes $\Lambda = 0$ [Watt, 1997].

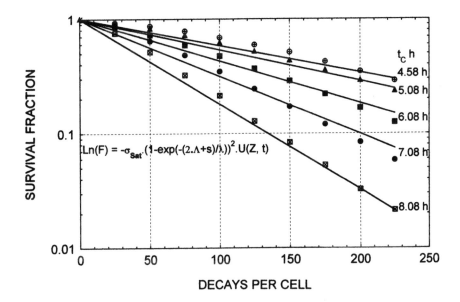

Figure 6.3(a). *Survival fractions, calculated from equation 6.3b, are shown as solid lines and are compared with the experimental data of Hofer et al., [1992] and Hofer and Bao [1995], for synchronized CHO cells irradiated by ^{125}I Auger electrons. The times into the cell cycle, t_C, indicated represent 0.5, 1, 2, 3, and 4 hours, respectively, into S phase. The mean deviation of the calculated points from the experimental values is 8.7%. [Reprinted with permission of Nuclear Technology Publishing, Ashford, Kent, England.]*

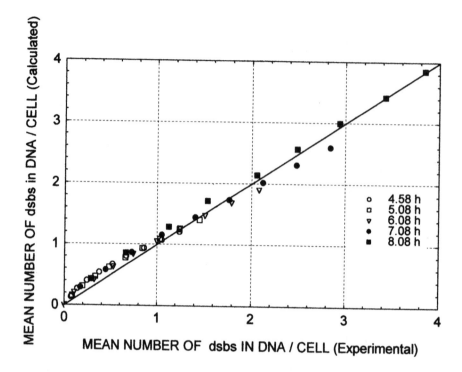

Figure 6.3(b). *Information on the mean number of dsbs in DNA, induced per cell by* ^{125}I *Auger electrons, which remain unrepaired, are extracted from the observed survival fractions shown in Figure 6.3(a) and compared with the values calculated using equations 6.3b and 6.5. The mean deviation of the 50 individual points is 6.9%.*

As each type of cell line has different characteristics with regard to size, radiosensitivity, and repair, it is difficult to design unambiguously clear tests of the theoretical model without the availability of a sufficiently extensive range of data endpoints on one cell line measured in the same, or closely collaborating, laboratories. Some confidence is obtained in the knowledge that the magnitudes of the key biophysicochemical quantities can be derived by other independent methods, for comparison. A useful test, the one adopted here, is to explore the ability of the model to describe simultaneously diverse measurements obtained with different radiation types delivered under different conditions. Independent measurement of the yields of dsbs per cell, shown in Figure 6.3(b) for Auger electron irradiations, would provide a critical test of the proposed damage mechanisms. At present the accuracy of the method used here is limited mainly by the uncertainty in the dynamics of cell repair mechanisms; i.e., in the validity of the assumptions made that lead to the format of the unrepair function, $U(Z,t)$; in the time dependence of the radiosensitivity, $Z(t)$; in the cell cycle and its influence on the $U(Z,t)$ for each cell line. Indeed, caution should be exercised in the use of a mean repair time as its existence as a meaningful quantity has yet to be proved. The shape of the survival curve is highly sensitive to the value of $U(Z,t)$ because it appears in the exponent of the formula for the surviving fraction. Finally, the parameters implicit in equation 6.3a permit the simplified unification of data, enable predictions to be made on the significance of trends in damage quantities with particle type, and can offer an explanation of unusual radiation effects.

Cross Section for Production of ssbs in the DNA

By analogy with equation 6.3b, σ_{ssb}, the cross section for production of an ssb in the DNA by the indirect action of a diffusing radical from a line source may be expressed as

$$\sigma_{ssb} = (\sigma_{g,DNA} \cdot 2 \cdot n_0 \cdot \frac{\Lambda}{d}) \cdot (1 - \exp(-2 \cdot \frac{\Lambda}{\lambda})) \tag{6.7}$$

Alternatively, σ_{ssb} can be related to the fundamental physical and chemical kinetics through the relation:

$$\sigma_{ssb} = \frac{\sigma_r \cdot v_r \cdot t_r}{\lambda_r} \cdot 10^{-1} \mu m^2 \tag{6.8}$$

where σ_r (μm^2) is the cross section for interaction with the DNA structure, of radicals diffusing from a line source with a mean velocity, v_r (cm/s) and a mean lifetime, t_r (s) in the intracellular fluid. $\sigma_r \cdot v_r$ is the chemical reaction rate constant. λ_r (nm) is the mean free path for radical production, taken here as the reciprocal of twice the ionization rate. The numerical factor of 10^{-1} converts the cross section to μm^2. By comparing equations 6.7 and 6.8 and substituting the required basic quantities, determined by independent methods, the value of the radical reaction rate can be estimated. Thus, taking $\lambda_r = 1.5$ nm for hydroxyl radicals gives the reaction rate as ~3×10^{-7} cm^3 per second per DNA segment, which corresponds to the correct order for the reaction rates ($\sigma_r \cdot v_r$) in mammalian DNA of 10^{12} cm^3/mole per second, reported for example by Chatterjee [Chatterjee and Holley, 1991], thereby suggesting an encouraging degree of internal consistency in the model.

Heavy Particle and Neutron Therapy

Because the single-track method of quantifying damage mechanisms differs radically from the conventional energy-based system currently used in radiation protection, it is of interest to consider the consequences in applications where the production of optimum damage is important. As, for example, in heavy particle therapy, with the proposed system, the damage is determined by the numbers of pairs of interactions (with a 2-nm spacing) that correlate with the strands in the double-stranded DNA to cause a dsb. The damage is thought to be caused by single one-dimensional line tracks of charged particles that act independently of the energy transfer in the collisions. Along the individual tracks, direct and indirect action will compete with a net effectiveness controlled by the number of double-stranded DNA segments at risk and by the efficiency term, $(1 - e^{-(2\Lambda + s)/\lambda})^2$ for induction of dsbs.

Much discussion has ensued on the optimum type of accelerated ion to be used in heavy particle therapy [Böhne, 1992; Kraft et al., 1992]. From the mechanisms proposed here, the maximum RBE will always occur when the mean free path for linear primary ionization of the charged particle is uniquely equal to ~2 nm in the cell nucleus. (The corresponding LETs are multivalued; viz., p, 75 keV/μm; α, 125 keV/μm; ^{12}C, 217 keV/μm; ^{20}Ne, 254 keV/μm; ^{40}Ar, 340 keV/μm.) Consequently, the biophysical therapeutic advantage is expected to be the same for any ion type having the requisite mean free path of 2 nm. There is an exception to this rule for protons (and proton recoils from neutrons) that, because of their short ranges (less than the mean chord distance through the cell nucleus), have maxima RBEs at ~47 keV/μm (λ = 4.6 nm) and 52 keV/μm (λ = 2.3 nm), respectively (see Figures 5.3 and 5.6). Thus, other factors to be considered are the ranges over which the 2-nm spacing can be sustained in the cell nucleus and the possible effects on surrounding healthy tissue of the delta-ray penumbra emitted by slow ions in the region of saturated damage (λ < 2 nm). Despite the large observed effects on surrounding healthy tissue, on a fluence basis the delta-ray *dose* contribution does not significantly alter the position of the maximum RBE and the effect goes unnoticed. In terms of effect cross sections, protons and neutrons are somewhat less damaging because of the range limitation of protons at λ ~ 2 nm, than are accelerated helium ions and other heavy ions with Z > 2. The latter ions with the same lambda should produce identical effects as measured by the effect cross section at the cellular level. At the tumor level, it is likely that an ion type can be selected which can sustain the optimum damage over the larger dimensions. Otherwise there seems to be no justification for building large machines to accelerate heavier ions as apparently no biophysical advantage will be gained, especially as the therapist has the facility to use "wiggling" techniques to spread the beam over the treatment area. Table 6.2 shows different tumor sizes over which optimum damage can be sustained.

Table 6.2. *Tumor size over which optimum damage can be sustained by the ion track.*

Accelerated Ion Type	Tumor Dimension
Protons	2.0 μm
Helium-4	16.7 μm
Boron-11	128.0 μm
Carbon-12	243.0 μm
Neon-20	750.0 μm
Argon-40	0.42 cm
Krypton-84	3.4 cm
Xenon-129	14.0 cm

Beam penetration to deep-seated tumors can be achieved more readily and economically with the lighter ions. In boron neutron capture therapy, it is interesting to note that the $^{10}B(n, \alpha)^7Li$ reaction products have a net lambda value of ~0.9 nm, which places them at the very favorable position just greater than the optimum saturation damage, free from delta-ray effects. For intra-cellular action, damage should be 100% efficient in making fractionation of treatment unnecessary, judging on the basis of the biophysical arguments.

Neutron therapy has two main disadvantages over charged heavy particles. First, neutrons are always fundamentally less damaging as many of the proton recoils generated at the most damaging energies (determined by their lambda value) of about 220 keV are less efficient than heavier charged particles at the same optimum lambda because they cannot penetrate the whole cell nucleus to interact with all the DNA segments available along the chord traversal. The second disadvantage with neutrons is that the goal posts do not remain fixed! As the neutron energy in the tumor is increased, the net effect of the proton recoils at higher energy become more efficient than those with the optimum lambda equal to 2 nm because of the stochastics involved. Optimum damage moves upwards in recoil energy as the neutron energy at the point of interest is increased.

Proposed Fluence-based System of Risk Control for Radiological Protection

On the basis of the proposed unified system of dosimetry an improved system of dose limitation based on fluence can be constructed for better risk control. When compared with the current regulatory system, anomalies emerge; e.g., the inappropriateness of using the same ICRP radiation weighting factor of 20 for fast neutrons, for heavy accelerated ions, and for natural alpha particles. For a fluence-based system, alpha particles have a maximum effect cross section three times smaller than the most damaging heavy particles. The maximum effect cross sections for neutrons are smaller than those for alpha particles because the most damaging recoils from neutrons have ranges less than the cell nucleus. The significant differences in effectiveness, as a function of photon energy for X- and gamma rays and fast electrons, per unit equilibrium fluence are appropriately quantified. In the case of neutron irradiations, the evaluation leads to a simple smooth effect curve that harmonizes with the histogram of radiation weighting factors recommended by ICRP, making the latter's step function unnecessary (Figure 6.4).

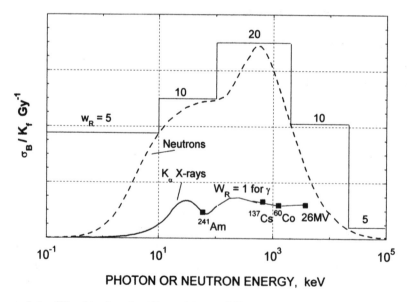

Figure 6.4. *The dose-based radiosensitivity of X- or gamma rays, and of neutrons, is expressed as a function of radiation energy. This can be directly compared with the ICRP radiation weighting factors shown as solid lines. The effect cross sections are obtained directly from equation 6.3b. [Reprinted with permission of Nuclear Technology Publishing, Ashford, Kent, England.]*

The ICRP risk coefficients are currently as follows:

1. For radiation workers, the risk corresponding to a dose limit of 20 mSv per year is determined from the cancer risk coefficient (R_{ICRP}) of 4×10^{-2} Sv^{-1} (called the "nominal probability coefficient" by ICRP) and is equal to $20 \times 10^{-3} \times 4 \times 10^{-2} = 8 \times 10^{-4}$ per year.

2. For the general population, the risk corresponds to a dose limit of 1 mSv per year. The cancer risk coefficient (R_{ICRP}) is 5×10^{-2} Sv^{-1} and the probability is $1 \times 10^{-3} \times 5 \times 10^{-2} = 5 \times 10^{-5}$ per year.

Risk factors, R_f per unit fluence, proposed in the new system, are related to the current ICRP risk factors by:

$$R_{f,\gamma} = R_{ICRP} \cdot \frac{K_{f.v_c}}{\sigma_{B,v_c}} \cdot Q_{v_c} \cdot \sigma_{B,\gamma} = 0.41 \times \sigma_{B,\gamma}$$

$$R_{f,n} = R_{ICRP} \cdot \frac{K_{f.n_c}}{\sigma_{B,n_c}} \cdot Q_{f,n_c} \cdot \sigma_{B,n} = 2.11 \times \sigma_{B,n} \qquad (6.9)$$

$$R_{Tot} = R_{f,\gamma} \cdot \Phi_\gamma + R_{f,n} \cdot \Phi_n$$

where Ks are kerma factors and Qs are quality factors for the reference radiations, subscript c. R_{Tot}, the net risk for a mixed field of photons and neutrons, is obtained by substitution into equation 6.9. Note that this equation permits the risk factors, R_f per unit fluence, to be obtained for any radiation type and energy spectrum through the use of the cross sections for the bioeffect, σ_B. If the risk factors are determined directly in terms of fluence, then the constant factors of 0.41 and 2.11 on the right-hand side of equation 6.9 should be unity. The results imply that the ICRP system underestimates the effect of gamma rays by 2.5 times and overestimates neutron effects by a factor of 2.

Implementation of a fluence-based system for radiation protection requires appropriate instrumentation. Devices to measure the absolute bioeffectiveness of any radiation type should, in principle, have a response that simulates that of the intracellular DNA and an output which yields the equivalent of the number of dsbs produced per unit fluence, independently of energy deposition. The latter response, in the absence of repair, is given by equation 6.3b with $U(Z,t)$ put equal to unity, and Figure 6.2, when expressed as cumulative probabilities. Such detectors are feasible.

7.

Conclusions

Our intention in the foregoing chapters was to show that absorbed dose, and the quantities linked to it, are not useful predictors of the effects of radiation. The evidence for this conclusion has been obtained on the one hand by analysis of published survival data for biological damage endpoints, mainly inactivation, and on the other hand from the evaluation of extensive research at low doses reported for a variety of radiation effects in living organisms.

From analyses of the cellular cross section data we conclude in Chapter 5 that lambda, the mean free path for linear primary ionization of the relevant charged particles, is a significantly better quality parameter than LET for correlating damage in a unified way for all radiation types. As lambda is not really connected to the energy transfer in a collision whereas LET self-evidently is, it follows that the amount of kinetic energy transferred in each collision plays no role in the production of radiation effects in mammalian cells (See *Relevance of Absorbed Dose for Specification of Radiation Effects* in Chapter 5). Consequently, as dose is equal to cema, the product of charged particle fluence and LET [Kellerer et al., 1992; Rossi and Kellerer, 1994], arguments against the validity of LET must apply equally to absorbed dose. A testable corollary to this is that individual delta rays from heavy particle tracks have negligible effect on the induction of damage. Evidence that such may indeed be the case is shown in Figures 5.6(a) and (b), but the conclusions are restricted by the rather large statistical errors. However, when aspects of the overall unified picture are taken into consideration; viz., the better correlation achieved with lambda; the (anomalous) excess damage attributed to delta-ray effects in the saturation region for slow heavy ions (Figures 5.3 and 5.5); the significantly lower degree of damage by incident photon and electron fields; the constancy of the observed cross sections for different ion types at the same lambda; the interpretation of the change in magnitude of the effect cross section "tails" for the slow heavy ions in the saturation region for different sizes of cell; and the variation in magnitude of the mean energy expended by heavy particles (Table 5.1) to inactivate a cell, then the conclusion becomes unequivocal. As a consequence, RBEs and modifying quality factors necessarily become redundant along with dose.

To a certain extent, our views are not too dissimilar from those of Katz (1994). These were summarized in a short commentary on dose in which he noted that the universal use of dose as a quantifying parameter in radiation is based entirely on the availability of the present types of measuring instruments. It is a poor basis for predicting or understanding the relationship between an irradiation and the resulting endpoint. Energy deposited is not the cause of an interaction; it is a secondary effect. The interaction is best described by fluence and cross section. If RBE is considered to be a correction factor to be applied to a prediction of response based on dose, it is found that its value can range from a very small fraction of unity to a very large multiple. It is apparent (concluded Katz) that, in general, dose is a useless predictor of response except in narrowly defined circumstances.

In the absence of energy transfer quantities, consideration of a new interpretation of the fundamental mechanism of radiation action is necessary. Figures 5.1 and 5.3 reveal several significant features, the most important of which is the correlation of a high proportion of the data points with two separate curves. One curve represents the damage arising from irradiation by electrons and X- and gamma rays; the other represents the damage arising from heavy charged particles. These two curves link at a mean spacing of ionizations of approximately 2 nm. The remaining data points lie on a third curve (the straight line of Figure 5.1) and are obtained from the irradiation of enzymes and viruses; i.e., organisms in which the replicating elements are single stranded. All those organisms which possess double-stranded DNA lie on the curves which show the point of inflexion at a position corresponding to the distance between the strands.

To accommodate these findings it is concluded that the radiation damage occurs when the linear spacing between the primary "ionizations" along individual tracks matches the interstrand spacing of the DNA to produce a dsb. This is a stochastical process which is adequately described by Poisson statistics and it is an optimum when the spacing is ~2 nm, at the onset of saturation damage, corresponding to a mean chord through the DNA helix. The radiation track must, so to speak, match the "template" of the strands of the DNA for an effective interaction to occur. Those interactions which occur at positions not so matched will have no effect, a situation that accounts for the irrelevance of energy transfer. There are on average ~15 pairs of strands at risk across the cell nucleus. The observed saturation cross section depends on the number of target DNA segments penetrated (determined by the particle projected range) and the interaction spacing along the relevant track.

A new model of radiation action which encompasses the foregoing comment is given in Chapter 6. It unifies the results for all radiation types and includes allowance for indirect and direct action, time, and whether the cells in their cycle are synchronized or asynchronized. The model is expected to serve in two ways: as a predictor of cell survival for any irradiation conditions and as the response function for instrumentation to be used in the proposed fluence-based system of dosimetry. Such instruments should be capable, at least in principle, of measuring the absolute bioeffectiveness without knowledge of the radiation field. As the model actually gives the yields of dsbs for any ionizing radiation, it will be fully testable in the near future with independent data which is becoming available for relevant DNA assays.

If, as we argue, absorbed dose does not give a sound basis for studying the effects of radiation (particularly at "low doses"), then one of the main tenets of radiation protection is effectively removed. This is the "linear, nonthreshold" hypothesis, which states that radiation-induced detriment, usually the induction of cancer, is linearly related to the dose absorbed, there being no threshold for the effect. The empirical evidence for and against this hypothesis has recently been discussed in great detail for low-LET external radiation [Kendall et al., 1992; Gilbert et al., 1993; Cardis et al., 1995; Schillaci, 1996; Mossman et al., 1996; Tubiana, 1996; Heidenreich et al., 1997; Rossi and Zaider, 1997]. The arguments will not be repeated here except to say that we believe that, to the extent that the measurement of low doses means anything, risks of health effects are either too small to be observed or are nonexistent for doses below 100 mSv.

The situation with regard to internal alpha emitters is very different. Of these, radium has received the most attention, and a report on some 2400 subjects has recently been published by Rowland [1994]. Because of its unequivocal evidence contradicting the "no-threshold" hypothesis, it is worth quoting from one of Rowland's conclusions:

> These results indicate that intake levels as large as 50 μCi of either radium isotope [i.e. ^{226}Ra or ^{228}Ra] produce bone changes that cannot be distinguished from changes sometimes appearing in unexposed individuals. Only when the intake levels are significantly larger do bone changes appear that indicate the presence of radium within the body.

In another part of this report, Rowland notes:

> No symptoms from internal radium have been recognized at levels lower than those associated with radium-induced malignancy. Radium levels 1000 times the natural ^{226}Ra found in all individuals apparently do little or no recognizable damage. These statements may suggest that a threshold exists for radium-induced malignancies; at least, they recognize that the available data demonstrate a steep dose response, with the risk dropping very rapidly for lower radium doses.

In a linked study, Thomas [1994] investigated doses and resultant radiation effects for a large percentage of the U.S. workers who were exposed to radium through their occupation of painting luminous dials. The data obtained were from females only because few males worked in this occupation. A log-normal analysis was carried out, and this proved to be a close fit to the data (46 sarcomas, 19 carcinomas). Most significantly, none of the 1400 subjects with average skeletal doses less than 10 Gy had developed any malignancy. This finding was in accord with that of Evans [1974], who found a "practical threshold" of 10 Gy for long-term effects of radium in the human body. (The methodology of converting from body burdens in µCi to skeletal doses in Gy is complex. Rowland [1994] gives full details of the various retention functions that have been used over the years.)

Information concerning the effects of plutonium in the human body is, inevitably, far more limited. In 1975, Voelz [Voelz, 1975] presented data on Atomic Energy Commission personnel who had been identified as having acquired internal plutonium depositions between 1957 and 1970. None of the 203 individuals had suffered any radiation-related pathological changes, not even those who had taken in over double the occupational permissible body burden (40 nanocuries). At about the same time, Durbin and her colleagues re-examined the survivors of the small group of patients of presumed short life expectancy who had been given plutonium in 1945-1946. Although the exact burdens were unknown, they have nevertheless been described as "high." Despite this, the examinations gave no indication that these individuals had experienced any detrimental effects from the plutonium injections [Rowland, 1994].

This study was subsequently extended to some 5000 workers whose systemic burden ranged from 7 to 230 µCi [Voelz et al., 1983]. The 452 observed deaths were significantly fewer than the 831 expected for all causes. The 107 deaths due to all malignant neoplasms were also significantly fewer than the 167 expected from these diseases. Based on plutonium depositions, the cohort was divided into "low" and "high" exposure groups, the division being made at 1 μCi per day. The results showed no increase in mortality in the high-exposure group as opposed to the low-exposure group. It would therefore appear that a threshold value of at least 230 μCi (i.e., at least six times the maximum permissible body burden) must exist for the induction of disease by plutonium.

The possible detrimental effects arising from the *inhalation* of plutonium are of much greater concern. They have therefore been the subject of many studies, with those conducted in the Pacific Northwest Laboratories being of greatest significance to radiation protection. The results reported by Sanders et al. [1993] are therefore summarized here.

A group of approximately 3000 rats were studied over their lifespan. As usual, these were divided into two groups, those that were sham-exposed or those that were exposed to an aerosol containing $^{239}PuO_2$. Mean lung doses given to the exposed rats ranged from 0.056 to 55 Gy. The incidence of malignant lung tumors was 1 in the 1052 control rats (<0.1%), none in the 1877 rats with doses <1 Gy (0%), and 39% in the 228 rats with doses >1 Gy. The threshold for the appearance of squamous cell carcinoma was 1.5 Gy and 3.1 Gy for adeno-carcinoma. Above this threshold, the incidence rose in an approximately quadratic way for the squamous cell carcinoma, and in an irregular way for the adeno-carcinoma.

These results are consistent with the numerous other studies cited by Sanders et al., [1993]. In none of these is there any evidence of a linear dose-response relationship with respect to lung tumor induction. It is of further interest to note that approximately 1 to 2 Gy was the mean lung dose required to ensure that all nuclei in the alveolar lung cells of a rat received at least one hit from an inhaled alpha emitter [Simmons, 1992].

The most serious hazard to the public-at-large is thought to arise from radon. This is the most widely occurring alpha emitter, and is found in high concentrations in several areas of the United Kingdom, the United States, and other countries. It also occurs in very high concentrations in underground mines, especially those in which uranium is extracted. Because of this, it too has been the subject of numerous studies, both in rodents and in humans. A detailed analysis of risks in rats exposed to radon has very recently been published by Gilbert et al. [1996].

The studies reported by Gilbert et al. [1996] were carried out over a long period of time and were very complex. To summarize these studies would be difficult, but one point is clear: evidence of a statistically significant excess cancer risk was limited to exposures of 80 working level months (WLMs) or greater.[1] The shape of the exposure response was investigated by fitting power functions of the form w^γ, where w denotes the cumulative exposure in WLM. Clearly, $\gamma = 1$ would correspond to a linear dose-effect response. When all the data were analyzed without considering exposure rate, the results were in reasonable agreement with a linear response function, especially if only those exposures below 1000 WLM were evaluated. However, when the exposure rate (or duration) was also considered, the results provided evidence that the power, γ, depended strongly on the exposure rate. For rats exposed at 100 WL the function was decidedly nonlinear, with γ estimated to be 1.5, although not so if only exposures below 1000 WLM were evaluated. Finally, there was no evidence of an inverse exposure-rate effect for cumulative exposures of < 1000 WLM.

Conversion from exposure in WLM to dose in gray is not a simple matter. It depends on (among other things) the age of the individual and the part of the lung considered. However, two independent sets of calculations [Hofmann, 1982; Haque and Al-Affan, 1988] have given a figure of approximately 10 mGy/WLM for the mid-lung sections of adults; a third set of calculations [Harley and Pasternak, 1982] gives a figure of about 5 mGy/WLM. If we accept the evidence of Gilbert et al., [1996] that there is a threshold of at least 80 WLM for the induction of cancer, this would convert to a value of about 0.8 Gy, a value similar to that found by Sanders et al., [1993] in his studies with plutonium.

Even more striking than the results obtained from rats are those obtained from uranium miners. In summarizing these, Duport [1994] has noted that the Standard Mortality Ratio or Relative Risk remains approximately constant over a wide range of exposures. This indicates that the increase in exposure has little effect on the incidence of lung cancer up to levels as high as 300 to 400 WLM; i.e., about 3 to 4 Gy. Not unreasonably, Duport suggests that these findings can be interpreted as practical thresholds for the induction of lung cancer by radon progeny.

Simultaneously, Birchall and James [1994] published their analysis of the large discrepancy between the results from the two different methods of calculating effective dose, E, per unit exposure, P. The first method (the

---

[1] Working level (WL) is defined as any combination of short-lived radon decay products in 1 liter of air resulting in the ultimate emission of 1.3×10^5 MeV of potential alpha-particle energy. This is equal to an activity concentration of 100 pCi/liter or 3.7×10^3 Bq/m^3. Working level month (WLM) is an exposure equivalent to 1 WL for 170 hours. Thus WL can be considered analogous to dose-rate and WLM analogous to total dose.

"dosimetric" method) involves the calculation of the dose to the cells believed to be at risk from inhaled radon and estimating the risk coefficients for lung cancer caused by uniform high dose-rate exposure to low LET radiation obtained from the Life Span Study. These are then modified by the appropriate weighting factors for alpha radiation in the lung. Using the dosimetric approach in conjunction with the recently published model of the human respiratory tract (ICRP 66, 1994) Birchall and James found a value for the conversion coefficient of 13.4 mSv per WLM. The second method, favored by the ICRP (ICRP 65, 1994), is based on the epidemiology of radon in mines and gives a value of about 5 mSv per WLM; i.e., nearly a factor of 3 lower. On the dosimetric basis, the implied risk is approximately 8.4×10^{-4} per WLM, whereas on the basis of the epidemiological data, the actual risk is not likely to be much higher than 2.8×10^{-4} per WLM. Based on these data, these authors concluded that the use of ICRP risk weighting factors results in an over-estimate of the lung cancer risk from radon progeny by a factor of 3. Furthermore, if this implication is accepted for radon progeny exposure, then it must also be considered for any alpha-emitting radionuclide because calculations of this type are not specific to radon progeny.

We trust that we have now led the reader of this monograph to the following conclusions:

1. Weighting coefficients and risk factors derived with their use are meaningless numbers when used for calculating radiation protection dosimetry.

2. To deny the existence of a threshold for the induction of cancer by radiation is to fly in the face of a large body of evidence supporting its existence.

3. The postulate of a linear response as a function of dose is, at best, a crude approximation and, at worst, an extremely expensive way to over-estimate risk.

4. For low-LET radiation, the fact that "dose" implies a homogeneous distribution of energy deposition means that it is a convenient surrogate for fluence down to values of about 1 c Gy. For high-LET (charged-particle) radiation, "dose" is a meaningless concept below values of a few gray.

In the light of the above, we wish to propose that a totally new system of measuring radiation be adopted. This system would be based on the fluence of the radiation, coupled with an estimation of the reciprocal mean free path of the primary ionization. Risk of detriment could then be established solely on the basis of these two measurements. Such a system would be the foundation of the radical reappraisal of radiation dosimetry that we and many other colleagues believe is overdue.

References

Chapter 1

There are no references for Chapter 1.

Chapter 2

Boag, John (1954). "The Relative Biological Efficiency of Different Ionizing Radiations." National Bureau of Standards Report 2946. Washington, D.C.

ICRP (1950). "International Recommendations on Radiological Protection." Report by the International Commission on Radiological Protection (1951). *British Journal of Radiology* 24: 46–53.

ICRP (1955). "Recommendations of the International Commission on Radiological Protection." *British Journal of Radiology* Supplement 6.

ICRP 6 (1962). "Recommendations of the International Commission on Radiological Protection." International Commission on Radiological Protection Publication 6. Oxford: Pergamon Press, 1964.

ICRP 26 (1977). "Recommendations of the International Commission on Radiological Protection," International Commission on Radiological Protection Publication 26. Oxford: Pergamon Press.

ICRU (1950). "International Recommendations on Radiological Units." Report by the International Commission on Radiological Units (1951). *British Journal of Radiology* 24: 54–56.

ICRU (1953). "Recommendations of the International Commission on Radiological Units." *Radiology* 62: 106–109.

ICRU 10a (1962). "Radiation Quantities and Units." International Commission on Radiation Units and Measurements Report 10a. Washington, D.C.: U.S. Government Printing Office.

ICRU 19 (1971). "Radiation Quantities and Units." International Commission on Radiation Units and Measurements Report 19. Washington, D.C.

Lea, Douglas E. (1946). *Actions of Radiations on Living Cells*. Cambridge, England: Cambridge University Press.

Chapter 3

BCRU Memo (1993). "Advice following ICRP Publication 60." Memorandum from the British Committee on Radiation Units and Measurements. *British Journal of Radiology* 66: 1201–1203 and in *Journal of Radiological Protection* 13: 71–73 and in *Radiation Protection Dosimetry* 46: 129–131.

BCRU Memo (1997). "Advice on the implications of the conversion coefficients for external radiations published in ICRP Publication 74." Memorandum from the British Committee on Radiation Units and Measurements. *Journal of Radiological Protection* 17: 201–204.

Bond, V. P. (1991). "When is a dose not a dose?" The Lauriston S. Taylor Lecture No. 15. National Council on Radiation Protection and Measurements. Bethesda, Md.

Bond, V. P. and M. N. Varma (1982). "A stochastic, weighted hit size theory of cellular radiobiological action." in Proceedings 8th Symposium on Microdosimetry. J. Booz and H. G. Ebert (eds.). Luxembourg: Commission of the European Communities, pp. 423–437.

Bond, V. P., V. Benary, and C. A. Sondhaus (1991). "A different perception of the linear, nonthreshold hypothesis for low-dose irradiation." *Proceedings of the National Academy of Sciences* 88: 8666–8670.

ICRP 26 (1977). "Recommendations of the International Commission on Radiological Protection." International Commission on Radiological Protection Publication 26. Oxford: Pergamon Press.

ICRP 60 (1990). "Recommendations of the International Commission on Radiological Protection." International Commission on Radiological Protection Publication 60. Oxford: Pergamon Press.

ICRP 74 (1996). "Conversion coefficients for use in radiological protection against external radiation." International Commission on Radiological Protection Publication 74. Oxford: Pergamon Press.

ICRU 36 (1983). "Microdosimetry." International Commission on Radiation Units and Measurements Report 36. Bethesda, Md.

Pelliccioni, M. and M. Silari (1993). "A critical view of radiation protection quantities for monitoring external radiation." *Journal of Radiological Protection* 13: 65–70.

Simmons, J. A. and S. R. Richards (1984). "Microdosimetry of alpha-irradiated lung." *Health Physics* 46: 607–616.

Simmons, J. A. and S. R. Richards (1989). "Microdosimetry of alpha-irradiated parenchymal lung." in *Low Dose Radiation: Biological Bases of Risk Assessment*. K. F. Baverstock and J. W. Stather (eds.). London: Taylor and Francis, pp. 312–324.

Chapter 4

Belli, M. et al., (1989). "RBE-LET relationship for the survival of V79 cells irradiated with low energy protons." *International Journal of Radiation Biology* 55: 93–104.

Blakely, E. A., C. A. Tobias, T. C. H. Yang, K. C. Smith, and J. T. Lyman (1979). "Inactivation of human kidney cells by high-energy monoenergetic heavy-ion beams." *Radiation Research* 80: 122–160.

Bond, V. P. and M. N. Varma (1981). "A stochastic weighted hit size theory of cellular radiobiological action." in Eighth Symposium in Micro-dosimetry. Jülich, Germany: Commission of the European Communities, pp. 423–438.

Chadwick, K. H. and H. P. Leenhouts (1973). "Chromosome aberrations and cell death." in 4th Symposium on Microdosimetry. EUR 5122, pp. 585–599.

Chadwick, K. H. and H. P. Leenhouts (1981). *The Molecular Theory of Radiation Biology*. Berlin: Springer-Verlag.

Chadwick, K. H., G. Moschini, and M. N. Varma (eds.) (1992). *Biophysical Modelling of Radiation Effects*. Bristol, Philadelphia and New York: Adam Hilger, 351.

Chatterjee, A. (1987). "Radiation chemistry." *Radiation Chemistry*. Farhataziz and M. Rodgers (eds.). VCH Publishers, 23.

Chatterjee, A. and W. R. Holley (1991). "Energy deposition mechanisms and biochemical aspects of DNA strand breaks by ionizing radiation." *International Journal of Quantum Chemistry* 39: 709.

Cox, R., J. Thacker, D. T. Goodhead, and R. J. Munson (1977). "Mutation and inactivation of mammalian cells by various ionising radiations." *Nature* 267: 425–427.

Curtis, S. B. (1986). "Lethal and potential lethal lesions induced by radiation—a unified repair model." *Radiation Research* 106: 252–270.

Curtis, S. B. (1987). "The cellular consequences of binary misrepair and linear fixation of initial biophysical damage." 8th International Congress of Radiation Research. E. M. Fielden, J. F. Fowler, J. H. Hendry, and D. Scott (eds.). Edinburgh, Scotland: Taylor and Francis, pp. 312–317.

Curtis, S. B. (1989). "The Katz cell-survival model and beams of heavy charged particles." Nuclear Tracks in Radiation Measurement. *International Journal of Radiation and Applied Instrumentation*, Part D, 16 (2/3): 97–103.

Datta, R., A. Cole, and S. Robinson (1976). "Use of track-end α-particles from ^{241}Am to study radiosensitive sites in CHO cells." *Radiation Research* 65: 139–151.

Dertinger, H. and H. Jung (1970). *Molecular Radiation Biology: The Action of Ionizing Radiation on Elementary Biological Objects*. New York, Heidelberg, Berlin: Springer-Verlag, 116.

Goodhead, D. T. (1989). "Relationships of radiation track structure to biological effect: a re-interpretation of the parameters of the Katz model." Nuclear Tracks in Radiation Measurement. *International Journal of Radiation and Applied Instrumentation*, Part D, 16 (2/3): 177–184.

Goodhead, D. T., J. Thacker, and R. Cox (1977). "The conflict between the biological effects of ultrasoft X-rays and microdosimetric measurements and application." 6th Symposium on Microdosimetry. Brussels: CEC.

Hall, E. J. (1988). *Radiobiology for the Radiobiologist.* 3rd ed. Philadelphia: J. P. Lippincott Co.

Harder, D. (1986). "Pairwise lesion interaction—extension and confirmation of Lea's model." 8th International Congress of Radiation Research. Edinburgh, Scotland: Taylor and Francis, pp. 318–324.

Harder, D. and P. Virsik-Peuckert (1984). "Kinetics of cell survival as predicted by the repair/interaction model." *British Journal of Cancer* 49 (Suppl. VI): 243–247.

Harder, D., R. Blohm, and M. Kessler (1988). "Restricted LET remains a good parameter of radiation quality." 6th Symposium on Neutron Dosimetry 23 (1/4): 79–82.

Harder, D., P. Virsik-Peuckert, and E. Bartels (1991). "Theory of pairwise lesion interaction." *Biophysical Modelling of Radiation Effects.* Padua, Italy: Adam Hilger, pp. 179–184.

Hill, C. K., F. M. Buonaguro, C. P. Myers, A. Han, and M. M. Elkind (1982). "Fission-spectrum neutrons at reduced dose-rates enhance neoplastic transformations." *Nature* 298: 67–69.

Holley, W. R., A. Chatterjee, and J. L. Magee (1990). "Production of DNA strand breaks by direct effects of heavy charged particles." *Radiation Research* 121: 161–168.

Humm, J., R. W. Howell, and D. V. Rao (1994). "Dosimetry of auger-electron emitting radionuclides: Report No. 3 of AAPM Nuclear Medicine Task Group No. 6." *Medical Physics* 21: 1901–1915.

ICRU 16 (1970). "Linear Energy Transfer." International Commission on Radiation Units and Measurements Report 16. Bethesda, Md.

ICRU 36 (1983). "Microdosimetry." International Committee on Radiation Units and Measurements Report 36, Bethesda, Md.

ICRU 37 (1984). "Stopping powers of electrons and positrons." International Commission on Radiation Units and Measurements Report 37. Bethesda, Md.

Katz, R. (1987). "Radiobiological modeling based on track structure." *Quantitative Mathematical Models in Radiobiology.* Schloss Rauisch-Holzhausen, Germany: Springer-Verlag, pp. 57–110.

Katz, R. (1994). "Commentary on 'Dose'." *Radiation Research* 137: 410–413.

Katz, R., S. C. Sharma, and M. Hoomayoonfar (1972). "The structure of particle tracks." *Topics in Radiation Dosimetry*, Supplement 1. F. H. Attix (ed.). New York and London: Academic Press, pp. 317–383.

Kellerer, A. M. and H. H. Rossi (1974). "The theory of dual radiation action." *Current Topics in Radiation Research.* M. Ebert and A. Howard (eds.). N. Holland, Amsterdam 8: 85–158.

References

Kellerer, A. M. and H. H. Rossi (1978). "A generalised formulation of dual radiation action." *Radiation Research* 75: 471–488.

Kiefer, J. (1987). "A repair fixation model based on classical enzyme kinetics." *Quantitative Mathematical Models in Radiation Biology*. Schloss Rauisch-Holzhausen, Germany: Springer-Verlag, pp. 171–179.

Lea, Douglas E. (1940). "Sizes of viruses and genes by radiation methods." *Nature* 146: 137–138.

Lea, Douglas E. (1955). *Actions of Radiations on Living Cells*. 2nd ed. Cambridge, England: Cambridge University Press, 416.

Neary, G. J. (1965). "Chromosome aberrations and the theory of RBE 1. General considerations." *International Journal of Radiation Biology* 9 (5): 477–502.

Perris, A., P. Pialoglou, A. A. Katsanos, and E. G. Sideris (1986). "Biological effectiveness of low energy protons I. Survival of Chinese hamster cells." *International Journal of Radiation Biology* 50: 1093–1102.

Rossi, H. H. and M. Zaider (1995). *Microdosimetry and its Applications*. Springer-Verlag, p. 315.

Sedlak, A. (1988). "Microdosimetric analysis of cell survival curves." *Acta Universitatis Carolinae: Medica* 34 (5/6): 291–333.

Simmons, J. A. (1992). "Absorbed dose—an irrelevant concept for irradiation with heavy charged particles?" *Journal of Radiological Protection* 12 (3): 173–179.

Skarsgard, L. D., D. A. Kihlmam, L. Parker, C. M. Pijara, and S. Richardson (1967). "Survival, chromosome abnormalities and recovery in heavy ion and X-irradiated mammalian cells." *Radiation Research*, Supplement 7: 208–221.

Sykes, C. E. and D. E. Watt (1989). "Interpretation of the increase in the frequency of neoplastic transformations observed for some ionising radiations at low dose rates." *International Journal of Radiation Biology* 55 (6): 925–942.

Todd, P. (1967). "Heavy ion irradiation of cultured human cells." *Radiation Research*, Supplement 7: 196–207.

Varma, M. N. and V. P. Bond (1987). "Hit-size effectiveness approach in biophysical modeling." *Quantitative Mathematical Models in Radiobiology*. Schloss Rauisch-Holzhausen, Germany: Springer-Verlag, pp. 119–124.

Waligórski, M. P. R. and P. Olko (1991). "Are neutron data suitable for making model predictions of radiation risk?" *Biophysical Modelling of Radiation Effects*. Padua, Italy: Adam Hilger, pp. 125–136.

Wambersie, A. (1990). "Radiobiological and clinical bases of particle therapy (review)." 2nd European Particle Accelerator Conference, Medical Satellite Meeting. Nice, France: Editions Frontieres S1–S3.

Watt, D. E. (1996). *Quantities for Dosimetry of Ionizing Radiations in Liquid Water*. 1st ed. Vol 1. London: Taylor and Francis.

Watt, D. E., I. A. M. Al-Affan, C. Z. Chen, and G. E. Thomas (1985). "Identification of biophysical mechanisms damage by ionizing radiation." *Radiation Protection Dosimetry* 13 (1/4): 285–294.

Watt, D. E., A. S. Alkharam, M. B. Child, and M.S. Salikin (1994). "Dose as a damage specifier in radiobiology for radiation protection." *Radiation Research* 139 (2): 249–251.

Watt, D. E., R. J. Cannell, and G. E. Thomas (1984). "Ion beams in quality determination for radiation protection." International Symposium on Three-day In Depth Review on the Nuclear Accelerator Impact in the Interdisciplinary Field. Padova, Italy: Laboratori Nazionali di Legnaro, pp. 267–282.

Wideroe, R. (1966). *Acta Radiologica* 4: 257.

Yamaguchi, H. and A. J. Waker (1982). "A resonance model for radiation action." 8th Symposium on Microdosimetry EUR 8395. Julich, Germany: Commission of the European Communities, Luxembourg, pp. 497–506.

Zermeno, A. and A. Cole (1969). "Radiosensitive structure of metaphase and interphase hamster cells as studied by low-voltage electron beam irradiation." *Radiation Research* 39: 669–684.

Zimmer, K. G. (1961). *Studies on Quantitative Radiation Biology*. 1st ed. Edinburgh and London: Oliver and Boyd, 124.

Selected Readings Chapter 4

Chapman, J. D. (1988). "Biophysical models of mammalian cell inactivation by radiation." *Radiation Biology in Cancer Research*, R. E. Meyn and H. R. Withers (eds.). New York: Raven Press, pp. 21–32.

Goodhead, D.T. (1991). "Biophysical modelling in radiation protection." *Biophysical Modelling of Radiation Effects*. Padua, Italy: Adam Hilger, Bristol, pp. 113–124.

Günther, K. and W. Schultz (1983). *Biophysical Theory of Radiation Action – A Treatise on Relative Biological Effectiveness*. Vol. 1. Berlin: Akademie-Verlag.

Kiefer, J. (1987). *Quantitative Mathematical Models in Radiation Biology*. Schloss Rauisch-Holzhausen, Germany: Springer-Verlag.

Kraft, G., M. Kramer, and M. Scholz (1992). "LET, track structure and models, a review." *Radiation and Environmental Biophysics* 31: 161–180.

Scholz, M. and G. Kraft (1994). "Calculation of survival in charged particle and neutron beams based on track structure." NIRS International Seminar on the Application of Heavy Ion Accelerators to Radiation Therapy of Cancer. Chiba, Japan: GSI–95–20.

Chapter 5

Alkharam, A. S. (1997a). "Specification of the quality of ionising radiations for unified dosimetry in radiobiology and radiological protection." School of Physics and Astronomy. St. Andrews: St. Andrews, Fife, Scotland, p. 236 and appendices.

Alkharam, A. S. (1997b). "Risk scaling factors from inactivation to chromosome aberrations, mutations and oncogenic transformations in mammalian cells." *Radiation Protection Dosimetry* 70 (1-4): 537–540.

Alper, T. (1979). *Cellular Radiobiology*. Cambridge, England: Cambridge University Press, p. 199.

Baltschukat, K. and G. Horneck (1991). "Responses to accelerated heavy ions of spores of bacillus subtilis of different repair capacity." *Radiation and Environmental Biophysics* 30: 87–103.

Baltschukat, K., G. Horneck, H. Bucker, R. Facius, and M. Schafer (1986). "Mutation induction in spores of bacillus subtilis by accelerated very heavy ions." *Radiation and Environmental Biophysics* 25: 183–187.

Baverstock, K. F. (1988). "DNA damage by Auger emitters." *DNA Damage by Auger Emitters*. Oxford, UK: Taylor and Francis.

Belli, M. et al. (1994). "Inactivation induced by deuterons of various LET in V79 cells." *Radiation Protection Dosimetry* 52 (1-4): 305–310.

Braby, L. A. (1995). "Quantitative description of particle interactions with biological samples." *Radiation Protection Dosimetry* 61 (1-3): 107–112.

Briden, P. E. (1988). "The Track Structure of Ionizing Particles and Their Application to Radiation Biophysics." Ph.D. Thesis. University of Westminster: London.

Bryant, P. E. (1985). "Enzymatic restriction of mammalian cell DNA: evidence for double-strand breaks as potentially lethal lesions." *International Journal of Radiation Biology* 48: 55–60.

Chadwick, K. H. and H. P. Leenhouts (1973). "Chromosome aberrations and cell death." 4th Symposium on Microdosimetry. EUR 5122, pp. 585–599.

Chadwick, K. H. and H. P. Leenhouts (1981). *The Molecular Theory of Radiation Biology*. Berlin: Springer-Verlag.

Charlton, D. E., D. T. Goodhead, W. E. Wilson, and H. G. Paretzke (1985). *Energy Deposition in Cylindrical Volumes: (a) Protons (b) Alpha particles*. MRC Radiobioloy Unit: Chilton, Didcot.

Chatterjee, A. and W. R. Holley (1991). "Energy deposition mechanisms and biochemical aspects of DNA strand breaks by ionizing radiation." *International Journal of Quantum Chemistry* 39: 709.

Cole, A., W. G. Cooper, F. Shonka, P. M. Corry, R. M. Humphrey, and A. T. Ansevin (1974). "DNA scission in hamster cells and isolated nuclei studied by low-voltage electron beam irradiation." *Radiation Research* 60: 1–33.

Cox, R., J. Thacker, D. T. Goodhead, and R. J. Munson, (1977). "Mutation and inactivation of mammalian cells by various ionising radiations." *Nature* 267: 425–427.

Datta, R., A. Cole, and S. Robinson (1976). "Use of track-end α-particles from ^{241}Am to study radiosensitive sites in CHO cells." *Radiation Research* 65: 139–151.

Dertinger, H. and H. Jung (1970). *Molecular Radiation Biology: The Action of Ionizing Radiation on Elementary Biological Objects*. New York, Heidelberg, Berlin: Springer-Verlag, 116.

Edwards, A. A. (1994). *The Induction of Chromosomal Changes in Human and Rodent cells by Accelerated Charged Particles: Early and Late Effects*. Oxford: National Radiological Protection Board: Didcot.

Folkard, M. et al. (1995). "Conventional and microbeam studies using low-energy charged particles relevant to risk assessment and the mechanisms of radiation action." *Radiation Protection Dosimetry* 61 (1-3): 215–218.

Frankenberg, D. (1994). "Repair of DNA double-strand breaks and its effect on RBE." *Advances in Space Research* 14: 235–248.

Hall, E. J. (1994). *Radiobiology for the Radiologist*. 4th ed. Philadelphia: J. B. Lippincott Company.

Harder, D. (1986). "Pairwise lesion interaction—extension and confirmation of Lea's model." 8th International Congress of Radiation Research. Edinburgh, Scotland: Taylor and Francis, pp. 318–324.

ICRU 36 (1983). "Microdosimetry." International Committee on Radiation Units and Measurements Report 36, Bethesda, Md.

Jagger, J. and E. C. Pollard (1956). "Inactivation of influenza A virus by fast charged particles." *Radiation Research* 4: 1–8.

Kellerer, A. M. and H. H. Rossi (1978). "A generalised formulation of dual radiation action." *Radiation Research* 75: 471–488.

Kellerer, A. M., K. Hahn, and H. H. Rossi (1992). "Intermediate Dosimetric Quantities." *Radiation Research* 130: 15–25.

Kiefer, J., S. Rase, E. Schneider, H. Stratten, G. Kraft, and H. Liesem (1982). "Heavy ion effects on yeast cells: induction of cerevise-resistant mutants." *International Journal of Radiation Biology* 42 (6): 591–600.

Kraft, G. (1987). "Radiobiological effects of very heavy ions: inactivation, induction of chromosome aberrations and strand breaks." *Nuclear Science Applications* 3 (1): 1–28.

Lea, Douglas E. (1955). *Actions of Radiations on Living Cells*. 2nd ed. Cambridge, England: Cambridge University Press, 416.

Neary, G. J. (1965). "Chromosome aberrations and the theory of RBE 1. General considerations." *International Journal of Radiation Biology* 9 (5): 477–502.

Perris, A., P. Pialoglou, A. A. Katsanos, and E. G. Sideris (1986). "Biological effectiveness of low energy protons. I. Survival of Chinese hamster cells." *International Journal of Radiation Biology* 50: 1093–1102.

Reist, H. W. et al. (1995). "A new method to study low dose radiation damage." *Radiation Protection Dosimetry* 61 (1-3): 221–224.

Rydberg, B. (1996). "Clusters of DNA damage induced by ionizing radiation: formation of short DNA fragments. ii. Experimental detection." *Radiation Research* 145: 200–209.

Thacker, J., A. Stretch, and M. A. Stephens (1979). "Mutation and inactivation of cultured mammalian cells exposed to beams of accelerated ions. II. Chinese hamster V79 cells." *International Journal of Radiation Biology* 36: 137–148.

References

Watt, D. E. (1989). "On absolute biological effectiveness and unified dosimetry." *Journal of Radiological Protection* 9 (1): 33–49.

Watt, D. E. (1996). *Quantities for Dosimetry of Ionizing Radiations in Liquid Water*. 1st ed. Vol 1. London: Taylor and Francis, 430.

Watt, D. E., R. J. Cannell, and G. E. Thomas (1984). "Ion beams in quality determination for radiation protection." International Symposium on Three-day In Depth Review on the Nuclear Accelerator Impact in the Interdisciplinary Field. Padova, Italy: Laboratori Nazionali di Legnaro, pp. 267–282.

Watt, D. E., I. A. M. Al-Affan, C. Z. Chen, and G. E. Thomas (1985). "Identification of biophysical mechanisms damage by ionizing radiation." *Radiation Protection Dosimetry* 13 (1/4): 285–294.

Wilson, D. and E. C. Pollard (1958). "Inactivation of Newcastle disease virus by ionizing radiation." *Radiation Research* 8: 131–148.

Yamaguchi, H. and A. J. Waker (1982). "A resonance model for radiation action." 8th Symposium on Microdosimetry EUR 8395. Julich, Germany: Commission of the European Communities, Luxembourg, 497–506.

Source of Data for Figures

Figure 5.1

Adams, G. E., E. M. Fielden, C. Hardy, B. C. Miller, I. J. Stratford, and C. Williamson (1981). "Radiosensitization of hypoxic mammalian cells in vitro by some 5-substituted-4-introimidazoles." *International Journal of Radiation Biology* 40: 153–161.

Barendsen, G. W., C. J. Koot, R. R. van Kersen, D. K. Bewley, S. B. Field, and C. J. Parnell (1966). "The effect of oxygen on impairment of the proliferative capacity of human cells in culture by ionizing radiations of different LET." *International Journal of Radiation Biology* 10: 317–327.

Belli, M., D. T. Goodhead, F. Ianzini, G. Simone, and M. A. Taboccini (1992). "Direct comparison of biological effectiveness of proton and alpha-particles of the same LET. II. Mutation induction at the HPRT locus in V79 cells." *International Journal of Radiation Biology* 61: 625–629.

Ben-Hur, E., M. M. Elkind, and B. V. Bronk (1974). "Thermally enhanced radioresponse of cultured Chinese hamster cells: inhibition of repair of sub-lethal damage and enhancement of lethal damage." *Radiation Research* 58: 38–51.

Bettega, D., et al. (1979). "Relative biological effectiveness for protons of energies up to 31 MeV." *Radiation Research* 77: 85–97.

Bird, R. P. and H. J. Burki (1975). "Survival of synchronised Chinese hamster cells exposed to radiation of different LET." *International Journal of Radiation Biology* 27: 105–120.

Blakely, E. A., C. A. Tobias, T. C. H. Yang, K. C. Smith, and J. T. Lyman (1979). "Inactivation of human kidney cells by high-energy mono-energetic heavy-ion beams." *Radiation Research* 80: 122–160.

Brustad, T. (1960). "Study of the radiosensitivity of dry preparations of lysozyme, trypsin and deoxyribonuclease exposed to accelerated nuclei of hydrogen, helium, carbon, oxygen and neon." *Radiation Research*, Supplement 2: 65–72.

Brustad, T. (1967). "Inactivation at various temperatures of the esterase activity of dried trypsin by radiations of different LET." *Radiation Research*, Supplement 7: 74–86.

Burki, H. J. and A. V. Carrano (1973). "Relative radiosensitivities of tetraploid and diploid Chinese hamster cells in culture exposed to ionizing radiation." *Mutation Research* 17: 277–282.

Chapman, J. D., R. G. Webb, and J. Borsa (1971). "Radiosensitization of mammalian cells by p-nitroacetophenone. I. Characterization in asynchronous and synchronous populations." *International Journal of Radiation Biology* 19: 561–573.

Cox, R., J. Thacker, and D. T. Goodhead (1971). "Inactivation and mutation of cultured mammalian cells by aluminium characteristic ultra-soft x-rays. II. Dose response of Chinese hamster and human diploid cells to aluminium x-rays and radiations of different LET." *International Journal of Radiation Biology* 31: 561–576.

Cox, R., J. Thacker, D. T. Goodhead, and R. J. Munson (1977). "Mutation and inactivation of mammalian cells by various ionising radiations." *Nature* 267: 425–427.

Dolphin, G. W. and F. Hutchinson (1960). "The action of fast carbon and heavier ions on biological material. I. The inactivation of dried enzymes." *Radiation Research* 13: 403–414.

Elkind, M. M. and H. Sutton (1959). "X-ray damage and recovery in mammalian cells in culture." *Nature* 184: 1293–1295.

Fluke, D. J., T. Brustad, and A. C. Birge (1960). "Inactivation of dried T1 bacteriophage by helium ions, carbon ions and oxygen ions: comparison of effects for tracks of various ion density." *Radiation Research* 13: 788–808.

Goodhead, D. T. and J. Thacker (1977). "Inactivation and mutation of cultured mammalian cells by aluminium characteristic ultrasoft x-rays: I. Properties of aluminium x-rays and preliminary experiments with Chinese hamster cells." *International Journal of Radiation Biology* 31: 541–559.

Goodhead, D. T., J. Thacker, and R. Cox (1977). "The conflict between the biological effects of ultrasoft X-rays and microdosimetric measurements and application." 6th Symposium on Microdosimetry. Brussels: CEC.

Hall, E. J., W. Gross, R. F. Dvorak, A. M. Kellerer, and H. H. Rossi (1972). "Survival curves and age response functions for Chinese hamster cells exposed to X-rays or high LET alpha particles." *Radiation Research* 52: 88–98.

References

Hall, E. J., A. M. Kellerer, H. H. Rossi, and Y. P. Lam (1978). "The relative biological effectiveness of 160 MeV protons." *International Journal of Radiation Oncology, Biology and Physics* 4: 1009–1013.

Han, A. and M. M. Elkind (1977). "Additive action of ionizing and non-ionizing radiations." *International Journal of Radiation Biology* 31: 275–282.

Harris, J. W., J. A. Power, and C. J. Koch (1975). "Radiosensitization of hypoxic mammalian cells by diamide. I. Effects of experimental conditions on survival." *Radiation Research* 64: 270–280.

Koch, C. J. and J. Kruuv (1971). "The effect of extreme hypoxia on recovery after irradiation by synchronised mammalian cells." *Radiation Research* 48: 74–85.

Kraft, G., et al. (1984). "HZE effects on mammalian cells." *Advances in Space Research* 4: 219–226.

McNally, N. J. (1976). "The effect of a change in radiation quality on the ability of electron affinic sensitizers to sensitize hypoxic cells." *International Journal of Radiation Biology* 29: 191–196.

Ngo, F. Q. H., A. Han, and M. M. Elkind (1977). "On the repair of sublethal damage in V79 Chinese hamster cells resulting from irradiation with fast neutrons or fast neutrons combined with X-rays." *International Journal of Radiation Biology* 32: 507–511.

Ngo, F. Q. H., E. A. Blakely, and C. A. Tobias (1981). "Sequential exposures of mammalian cells to low energy transfer and high linear energy transfer radiations: I. Lethal effects following x-ray and neon ion irradiation." *Radiation Research* 87: 59–78.

Perris, A., P. Pialoglou, A. A. Katsanos, and E. G. Sideris (1986). "Biological effectiveness of low energy protons. I. Survival of Chinese hamster cells." *International Journal of Radiation Biology* 50: 1093–1102.

Piro, A. J., C. C. Taylor, and J. A. Belli (1975). "Interaction between radiation and drug damage in mammalian cells. I. Delayed expression of actinomycin D/X-ray effects in exponential and plateau phase cells." *Radiation Research* 63: 346–362.

Raaphorst, G. P. and J. Kruuv (1978). "The radiation response of cultured mammalian V79-S171 cells exposed to a wide concentration range of sulphare salt solution." *International Journal of Radiation Biology* 33: 173–183.

Schambra, P. E. and F. Hutchinson (1964). "The action of fast heavy ions on biological material II. Effects on T1 and fX-174 bacteriophage and double-stranded and single-stranded DNA." *Radiation Research* 23: 514–526.

Schlag, H., K. F. Weibezahn, and C. Lucke-Huhle (1978). "Negative p_{ion} irradiation of mammalian cells. II. A comparative analysis of cell cycle progression after exposure to pi mesons and Co-60 gamma rays." *International Journal of Radiation Biology* 33: 1–10.

Skarsgard, L. D., D. A. Kihlmam, L. Parker, C. M. Pijara, and S. Richardson (1967). "Survival, chromosome abnormalities and recovery in heavy ion and X-irradiated mammalian cells." *Radiation Research*, Supplement 7: 208–221.

Stoll, U., A. Schmidt, E. Schneider, and J. Kiefer (1995). "Killing and mutation of Chinese hamster V79 cells exposed to oxygen and neon ions." *Radiation Research* 142: 288–294.

Thacker, J., A. Stretch, and M. A. Stephens (1979). "Mutation and inactivation of cultured mammalian cells exposed to beams of accelerated ions. II. Chinese hamster V79 cells." *International Journal of Radiation Biology* 36: 137–148.

Todd, P. W. (1975). "Heavy ion irradiation of human and Chinese hamster cells in vitro." *Radiation Research* 61: 288–297.

Utsumi, H. and M. M. Elkind (1979). "Potentially lethal damage qualitative differences between ionizing and non-ionizing radiation and implications for 'single-hit' killing." *International Journal of Radiation Biology* 35: 373–380.

Wainson, A. A., M. F. Lomanov, N. L. Shmakova, S. I. Blokhin, and S. P. Jarmenko (1972). "The RBE of accelerated protons in different parts of the Bragg curve." *British Journal of Radiology* 45: 525–529.

Waldren, C. A. and I. Rasko (1978). "Caffein enhancement of X-ray killing in cultured human and rodent cells." *Radiation Research* 73: 95–110.

Yatagai, F., S. Kitayama, and A. Mutsuyama (1979). "Inactivation of phage φX-174 by accelerated ions." *Radiation Research* 77: 250–258.

Figures 5.2 and 5.3

Belli, M. et al. (1989). "RBE-LET relationship for the survival of V79 cells irradiated with low energy protons." *International Journal of Radiation Biology* 55: 93–104.

Belli, M., F. Cera et al. (1993). "Inactivation and mutation induction in V79 cells by low energy protons: re-evaluation of the results at the LNL facility." *International Journal of Radiation Biology* 63: 331–337.

Belli, M. et al. (1994). "Inactivation induced by deuterons of various LET in V79 cells." *Radiation Protection Dosimetry* 52 (1-4): 305–310.

Bird, R. P. and H. J. Burki (1975). "Survival of synchronised Chinese hamster cells exposed to radiation of different LET." *International Journal of Radiation Biology* 27: 105–120.

Cherubini, R. (1994). "Molecular and cellular effectiveness of charged particles (light and heavy ions) and neutrons. Project 1." Laboratori Nazionali di Legnaro, Instituto Nazionale di Fisica Nucleare: Legnaro, Italy.

Cox, R., J. Thacker, D. T. Goodhead, and R. J. Munson, (1977). "Mutation and inactivation of mammalian cells by various ionising radiations." *Nature* 267: 425–427.

Folkard, M., K. M. Prise, B. Vojnovic, S. Davies, M. J. Roper, and B. D. Michaels (1989). "The irradiation of V79 mammalian cells by protons with energies below 2 MeV. Part 1: Experimental arrangement and measurement of cell survival." *International Journal of Radiation Biology* 56: 221–237.

Folkard, M., K. M. Prise et al. (1996). "Inactivation of V79 by low-energy protons, deuterons and helium-3 ions." *International Journal of Radiation Biology* 69: 729–738.

Fox, J. C. and N. J. McNally (1988). "Cell survival and DNA double-strand break repair following X-ray or neutron irradiation of V79." *International Journal of Radiation Biology* 54: 1021–1030.

Goodhead, D.T., M. Belli et al. (1992). "Direct comparison between protons and alpha-particles of the same LET: irradiation methods and inactivation of asynchronous V79, HeLa and C3H10T1/2 cells." *International Journal of Radiation Biology* 61: 611–624.

Hall, E. J., J. K. Novak et al. (1975). "RBE as a function of neutron energy. I. Experimental observations." *Radiation Research* 64: 245–255.

Hall, E. J., R. P. Bird et al. (1977). Biophysical studies with high-energy argon ions. 2. Determination of the relative biological effectiveness, the oxygen enhancement ratio and cell cycle response." *Radiation Research* 70: 469–479.

Hall, E. J., A. M. Kellerer, H. H. Rossi, and Y. P. Lam (1978). "The relative biological effectiveness of 160 MeV protons." *International Journal of Radiation Oncology, Biology and Physics* 4: 1009–1013.

Hall, E. J., A. M. Kellerer et al. (1982). "Dependence on neutron energy of the OER and RBE." *International Journal of Radiation Oncology, Biology and Physics* 8: 1567–1572.

Jenner, T. J., C. M. DeLara et al. (1993). "Induction and rejoining of DNA strand breaks in V79-4 mammalian cells following gamma and alpha irradiation." *International Journal of Radiation Biology* 64: 265–273.

Key, M. (1971). "Comparative effects of neutrons and x-rays on Chinese hamster cells." *Biophysical Aspects of Radiation Quality*. Vienna, Austria: International Atomic Enery Authority (IAEA), pp. 431–444.

Kraft, G., W. Kraft-Weyrather, H. Meister, G. Miltenburger, R. Roots, and H. Wulf (1982). "The influence of radiation quality on the biological effectiveness of heavy charged particles." 8th Symposium on Microdosimetry, EUR 8395. Julich, Germany: Commission of the European Communities, Luxembourg, pp. 743–753.

Kranert, T. and E. Schneider (1990). "Mutation induction in V79 Chinese hamster cells by very heavy ions." *International Journal of Radiation Biology* 58: 975–987.

Kranert, T., U. Stoll, E. Schneider, and J. Kiefer (1992). "Mutation induction in mammalian cells by very heavy ions." *Advances in Space Research* 12 (2): 111–118.

Lucke-Huhle, C., E. A. Blakely et al. (1979). "Drastic G2 arrest in mammalian cells after irradiation with heavy-ion beams." *Radiation Research* 79: 97–112.

Min, T., P. Cohn, and J. A. Simmons (1985). "The sensitivity of three diploid human cell lines to alpha-irradiation." Second Workshop on Lung Dosimetry. Cambridge University.

Munson, R. J., D. A. Bance, A. Stretch, and D. T. Goodhead (1979). "Mutation and inactivation of cultured mammalian cells exposed to beams of accelerated ions. I. Irradiation facilities and methods." *International Journal of Radiation Biology* 36: 127–136.

Ngo, F. Q. H., E. A. Blakely, and C. A. Tobias (1981). "Sequential exposures of mammalian cells to low energy transfer and high linear energy transfer radiations: I. Lethal effects following x-ray and neon ion irradiation." *Radiation Research* 87: 59–78.

Perris, A., P. Pialoglou, A. A. Katsanos, and E. G. Sideris (1986). "Biological effectiveness of low energy protons. I. Survival of Chinese hamster cells." *International Journal of Radiation Biology* 50: 1093–1102.

Prise, K. M., S. Davies et al. (1987). "The relationship between radiation-induced DNA double-strand breaks and cell kill in hamster V79 fibroblasts irradiated with 250 kVp X-rays, 2.3 MeV neutrons or ^{238}Pu alpha-particles." *International Journal of Radiation Biology* 52: 893–902.

Prise, K. M. and M. Folkard (1990). "The irradiation of V79 mammalian cells by protons with energies below 2 MeV. Part II. Measurement of oxygen enhancement ratios and DNA damage." *International Journal of Radiation Biology* 58: 261–277.

Railton, R., R. C. Lawson et al. (1973). "Neutron spectrum dependence of RBE and OER values." *International Journal of Radiation Biology* 23: 509–518.

Raju, M. R., Y. Eisen, S. Carpenter, and W. C. Inkret (1991). "Radiobiology of alpha-particles. III. Cell inactivation by alpha-particle traversals of the cell nucleus." *Radiation Research* 128: 204–209.

Schlag, H. and C. Lucke-Huhle (1981). "The influence of ionization density on the DNA synthetic phase and survival of irradiated mammalian cells." *International Journal of Radiation Biology* 40: 75–85.

Simmons, J. A., P. Cohn, and T. Min (1996). "Survival and yields of chromosome aberrations in hamster and human lung cells irradiated by alpha particles." *Radiation Research* 145: 174–180.

Sinclair, W. K. (1985). "Experimental RBE values of high LET radiations at low doses and the implications for quality factor assignment." *Radiation Protection Dosimetry* 13 (1-4): 319–326.

Stenerlow, B., E. Blomquist et al. (1996). "Re-joining of DNA double-strand breaks induced by accelerated nitrogen ions." *International Journal of Radiation Biology* 70: 413–420.

Thacker, J., A. Stretch, and M. A. Stephens (1979). "Mutation and inactivation of cultured mammalian cells exposed to beams of accelerated ions. II. Chinese hamster V79 cells." *International Journal of Radiation Biology* 36: 137–148.

Thacker, J., A. Stretch, and D. T. Goodhead (1982). "The mutagenicity of alpha-particles from plutonium-238." *Radiation Research* 92: 343–352.

Todd, P. W. (1975). "Heavy ion irradiation of human and Chinese hamster cells in vitro." *Radiation Research* 61: 288–297.

References

Tolkendorf, E. and K. Eichhorn (1983). "Effect of ionizing radiation of different linear energy transfer on the induction of cellular death and of chromosome aberrations in cells of the Chinese hamster." *Studia Biophysica* 95: 43–56.

Wainson, A. A., M. F. Lomanov, N. L. Shmakova, S. I. Blokhin, and S. P. Jarmenko (1972). "The RBE of accelerated protons in different parts of the Bragg curve." *British Journal of Radiology* 45: 525–529.

Weber, K. J. and M. Flentje (1993). "Lethality of heavy-ion induced DNA double-strand breaks in mammalian cells." *International Journal of Radiation Biology* 64: 169–178.

Wouters, B. G., G. K. Y. Lam et al. (1996). "Measurements of relative biological effectiveness of the 70 MeV proton beam at TRIUMF using Chinese hamster V79 cells and the high precision cell sorter assay." *Radiation Research* 146: 159–170.

Wulf, H., W. Kraft-Weyrather, H. G. Miltenburger, E. A. Blakely, C. A. Tobias, and G. Kraft (1985). "Heavy-ion effects on mammalian cells: inactivation measurements with different cell lines." *Radiation Research* 104: S122–S134.

Figure 5.5

Baltschukat, K., G. Horneck, H. Bucker, R. Facius, and M. Schafer (1986). "Mutation induction in spores of bacillus subtilis by accelerated very heavy ions." *Radiation and Environmental Biophysics* 25: 183–187.

Chapter 6

Alkharam, A. S. (1997). "Risk scaling factors from inactivation to chromosome aberrations, mutations and oncogenic transformations in mammalian cells." *Radiation Protection Dosimetry* 70 (1-4): 537–540.

Alkharam, A. S. and D. E. Watt (1997). A Radiobiological Data-base: Cellular Inactivation, Dicentrics, Mutations, Oncogenic Transformations, and DNA Strand Breaks. St. Andrews, Fife: University of St. Andrews.

Böhne, D. (1992). "Light ion accelerators for cancer therapy." *Radiation and Environmental Biophysics* 31: 205–218.

Charlton, D. E. and J. Booz (1987). "Local effects of Auger electron cascades." Eighth International Congress of Radiation Research. Edinburgh: Taylor and Francis, July, 1997, pp. 363–374.

Chatterjee, A. and W. R. Holley (1991). "Energy deposition mechanisms and biochemical aspects of DNA strand breaks by ionizing radiation." *International Journal of Quantum Chemistry* 39: 709.

Chilton, A. B. (1979). "Further comments on an alternative definition of fluence." *Health Physics* 36: 637–638.

Curtis, S. B. (1986). "Lethal and potentially lethal lesions induced by radiation—a unified repair model." *Radiation Research* 106: 252–270.

Grosswendt, Bernd (1997). Private Communication, Padua, Italy, June 10.

Hall, E. J. (1994). *Radiobiology for the Radiologist*. 4th ed. Philadelphia: J. B. Lippincott Company.

Harder, D. and P. Virsik-Peuckert (1984). "Kinetics of cell survival as predicted by the repair/interaction model." *British Journal of Cancer* 49 (Suppl. VI): 243–247.

Hofer, K. G. and S.-P. Bao (1995). "Low-LET and high-LET radiation action of ^{125}I decays in DNA: effect of cysteamine on micronucleus formation and cell killing." *Radiation Research* 141: 183–192.

Hofer, K. G., N. van Loon, M. H. Schneiderman, and D. E. Charlton (1992). "The paradoxical nature of DNA damage and cell death induced by ^{125}I decay." *Radiation Research* 130: 121–124.

Howell, R. W. and D. V. Rao (1996). "Bibliography: biophysical aspects of Auger processes (1964 to May, 1995)." Third International Symposium on Biophysical Aspects of Auger Processes. University of Lund, Sweden, August 24-25, 1995.

Humm, J., R. W. Howell, and D. V. Rao (1994). "Dosimetry of Auger-electron emitting radionuclides: Report No. 3 of AAPM Nuclear Medicine Task Group No. 6." *Medical Physics* 21: 1901–1915.

Kiefer, J. (1987). "A repair fixation model based on classical enzyme kinetics." *Quantitative Mathematical Models in Radiation Biology*. Schloss Rauisch-Holzhausen, Germany: Springer-Verlag, pp. 171–179.

Kraft, G., M. Kramer, and M. Scholz (1992). "LET, track structure and models." *Radiation Environmental Biophysics* 31 (3): 161–180.

Lea, Douglas E. (1955). *Actions of Radiations on Living Cells*. 2nd ed. Cambridge, England: Cambridge University Press, 416.

McDougall, I. C., A. S. Alkharam, G. E. Thomas, and D. E. Watt (1997). "A scintillation method for absolute dosimetry by simulation of the radiation response to mammalian cells." The IRPA Regional Symposium on Radiation Protection in Neighbouring Countries of Central Europe. Prague.

Sinclair, W. (1972). "Cell cycle dependence of the lethal radiation response in mammalian cells." *Current Topics in Radiation Research* 7: 264–285.

Steel, G.G. (1989). "The picture has changed since the 1980's." *International Journal of Radiation Biology* 56 (No. 5): 525–537.

Watt, D. E. (1996). *Quantities for Dosimetry of Ionizing Radiations in Liquid Water*. 1st ed. Vol. 1. London: Taylor and Francis, 430.

Watt, D. E. (1997). "A unified system of radiation bio-effectiveness and its consequences in practical application." *Radiation Protection Dosimetry* 70 (1-4): 529–536.

Watt, D. E. and A. S. Alkharam (1995). "A feasibility study of scintillator microdosemeters for measurement of the biological effectiveness of ionizing radiations." *Radiation Protection Dosimetry* 61 (1-3): 211–214.

Weber, D. E., K. F. Eckerman, L. T. Dillman, and J. C. Ryman (1989). *MIRD: Radionuclide Data and Decay Schemes*. New York: The Society of Nuclear Medicine, 447.

References

Chapter 7

Birchall, A. and A. C. James (1994). "Uncertainty analysis of the effective dose per unit exposure from radon progeny and implications for ICRP risk-weighting factors." *Radiation Protection Dosimetry* 53: 133–140.

Cardis, E., E. S. Gilbert, L. Carpenter, G. Howe, I. Kato, B. K. Armstrong, V. Beral, G. Cowper, A. Douglas, J. Fix, S. A. Fry, J. Kaldor, C. Lave, L. Salmon, P. G. Smith, G. L. Voelz, and L. D. Wiggs (1995). "Effects of low doses and low dose rates of external ionizing radiation: cancer mortality among nuclear industry workers in three countries." *Radiation Research* 142: 117–132.

Duport, P. (1994). "Radiation protection in uranium mining: two challenges." *Radiation Protection Dosimetry* 53: 13–19.

Evans, R. D. (1974). "Radium in man." *Health Physics* 27: 497–510.

Gilbert, E. S., D. L. Cragle, and L. D. Wiggs (1993). "Updated analyses of combined mortality data for workers at the Hanford Site, Oak Ridge National Laboratory and Rocky Flats Weapons Plant." *Radiation Research* 136: 408–421.

Gilbert, E. S., F. T. Cross, and G. E. Dagle (1996). "Analysis of lung tumour risks in rats exposed to radon." *Radiation Research* 145: 350–360.

Haque, A. K. M., and I. A. M. Al-Affan (1988). "Radiation dose to the lungs due to inhalation of alpha emitters." *Health Effects of Low Dose Ionising Radiation*. London: British Nuclear Energy Society.

Harley, N. H. and B. S. Pasternak (1982). "Environmental radon daughter alpha dose factors in a five-lobed human lung." *Health Physics* 42: 789–799.

Heidenreich, W. E., H. G. Paretzke, and P. Jacob (1997). "No evidence for increased tumor rates below 200 mSv in the atomic bomb survivor data. *Radiation and Environmental Biophysics* 36: 205–207.

Hofmann, W. (1982). "Cellular lung dosimetry for inhaled radon decay products as a base for radiation-induced lung cancer risk assessment. 1. Calculation of mean cellular doses." *Radiation and Environmental Biophysics* 20: 95–112.

ICRP 65 (1994). "Protection Against Radon-222 at Home and at Work." International Commission on Radiological Protection Publication 65. Ann. ICRP 23. Oxford: Pergamon Press.

ICRP 66 (1994). "Human Respiratory Tract Model for Radiological Protection." International Commission on Radiological Protection Publication 66. Ann. ICRP 24. Oxford: Pergamon Press.

Katz, R. (1994). Commentary on "Dose." *Radiation Research* 137: 410–413.

Kellerer, A. M., K. Hahn, and H. H. Rossi (1992). "Intermediate dosimetric quantities." *Radiation Research* 130: 15–25.

Kendall, G. M., C. R. Muirhead, B. H. MacGibbon, J. A. O'Hagan, A. J. Conquest, A. A. Goodill, B. K. Butland, T. P. Fell, D. A. Jackson, M. A. Webb, R. G. Haylock, J. M. Thomas, and T. J. Silk (1992). "Mortality and occupational exposure to radiation: first analysis of the National Registry for Radiation Workers." *British Medical Journal* 304: 220–225.

Mossman, K. L., M. Goldman, F. Massé, W. A. Mills, K. J. Schlager, and R. J. Vetter (1996). "Radiation risk in perspective: a position statement of the Health Physics Society." HPS Newsletter 24 (March 1996).

Rossi, H. H. and A. M. Kellerer (1994). "Why Cema?" ICRU News 6-8 (Dec. 1994).

Rossi, H. H. and M. Zaider (1997). Radiogenic lung cancer: the effects of low doses of low linear energy transfer (LET) radiation." *Radiation and Environmental Biophysics* 36: 85–88.

Rowland, R. E. (1994). *Radium in Humans; A Review of U.S. Studies*. Argonne National Laboratory.

Sanders, C. L., K. E. Lauhala, and K. E. McDonald, (1993). "Lifespan studies in rats exposed to $^{239}PuO_2$ aerosol. III. Survival and lung tumours." *International Journal of Radiation Biology* 64: 417–430.

Schillaci, M. E. (1996). Comments on "Effects of low doses and low dose rates of external ionizing radiation." *Radiation Research* 145: 647–651.

Simmons, J. A. (1992). "Absorbed dose—an irrelevant concept for irradiation with heavy charged particles?" *Journal of Radiological Protection* 12: 173–179.

Thomas, R. G. (1994). "The U. S. radium luminisers: a case for a policy of 'below regulatory concern.'" *Journal of Radiological Protection* 14:141–153.

Tubiana, M. (1996). "Effets cancérogènes des faibles doses du rayonnement ionisant." *Radioprotection* 31: 155–191.

Voelz, G. L. (1975). "What have we learned about plutonium from human data?" *Health Physics* 29: 551–561.

Voelz, G. L., G. S. Wilkinson, J. F. Acquavella, G. L. Tietjen, R. N. Brackbill, M. Reyes, and L. D. Wiggs (1983). "An update of epidemiological studies of plutonium workers." *Health Physics* 44: 493–503.

Index

About the Authors

Jack A. Simmons, Ph.D.

Dr. Jack Simmons received his Ph.D. degree in Radiation Physics from the Medical College of St. Bartholomew's Hospital, University of London. After two years as a Postdoctoral Fellow at the Johns Hopkins University in Baltimore, he returned to England to join a team of biologists and chemists at the Medical Research Council, Hammersmith Hospital in London. He subsequently moved into teaching, and since 1970 has been at the Polytechnic of Central London (now the University of Westminster). The title of Professor of Radiation Biophysics was conferred on him in 1995. Dr. Simmons' early research lay in the field of radiation-induced free radicals. However some of his results led him to believe that a measurement of the quantity of radiation absorbed (the "dose") was not adequate in understanding the observed effects. The advent of "microdosimetry," which considers the detailed distribution of energy deposition events, therefore greatly excited him. He consequently switched fields to apply this new concept to the problem of the effects of inhaled alpha particle emitters such as plutonium and radon in the lung. His findings have been published in many journals such as *Health Physics* and *Radiation Research* as well as in the Proceedings of several symposia in microdosimetry.

David E. Watt, Ph.D., D.Sc.

Dr. David Watt obtained his Ph.D. degree in nuclear physics at the University of Glasgow. In 1956, he joined the UK Atomic Weapons Research Establishment at Aldermaston, where he remained until 1961. As a Senior Scientific Officer, Dr. Watt was engaged in the design and development of high sensitivity detectors for absolute measurement of radionuclides resulting from nuclear fall-out. Then, joining a Radiobiological Research Unit in the nuclear power sector of the UK Atomic Energy Authority at Chapelcross, Dumfriesshire, Dr. Watt specialized in the study of the biophysical mechanisms of radiation damage, particularly for intermediate energy neutrons, with the objective of specifying radiation quality. In 1969, along with other members of the Unit at Chapelcross, he relocated within the Medical Faculty of the University of Dundee, Scotland, forming the new Department of Medical Biophysics to provide expertise in medical applications of ionizing radiations as well as postgraduate teaching. His main research explored methods of quantifying radiation effects for all radiation types into a unified system for absolute dosimetry. During this period he was appointed Reader in Radiation Biophysics.

In 1976 the University of Dundee conferred on him the degree of Doctor of Science for "studies in atomic and nuclear radiations." In 1980, he was awarded the Founders Medal of the UK Society for Radiological Protection "for contributions of distinction." In 1984 Dr. Watt transferred to the School of Physics and Astronomy at the University of St. Andrews where his main research objective was to find a method to unify the specification of radiation effects in therapy and protection. His current research interest, in conjunction with European colleagues, is to develop new types of detectors for implementation of the dosimetry system.

Dr. Watt has published over 150 papers in diverse radiation topics; a research monograph (with D. Ramsden), *High Sensitivity Counting Techniques* (Pergamon Press, 1961) and a book of tables, *Quantities for Dosimetry of Ionizing Radiations in Liquid Water* (Taylor & Francis, 1996).